METRIC CONVERSION TABLES AND FACTORS

BY THE EDITORS OF
MACHINERY'S HANDBOOK

INDUSTRIAL PRESS INC.
200 Madison Avenue, New York, N.Y. 10016

THE MACHINERY PUBLISHING COMPANY LTD.
Brighton, England

Library of Congress Cataloging in Publication Data

Main entry under title:

Metric conversion tables and factors.

1. Metric system — Conversion tables. I. Machinery's handbook.

QC94.M38 389′.152 73-8876

ISBN 0-8311-1103-8

METRIC CONVERSION FACTORS AND TABLES

TABLE OF CONTENTS

INTRODUCTION

The information presented here is intended for practical use in engineering and related technical fields, although other branches of industry should also find it useful. Emphasis is placed on the International System of Units (SI) which is the modernized metric system of measurement.

This emphasis is in keeping with the worldwide move to adopt SI, and the growing interest in this system in the U.S.A.

But the contents is not limited to conversions into SI units. Units from earlier metric systems are also given. The reason is that readers will meet with such units as kilogram-force, metric horsepower or bar and will need to know what the values are in SI or customary English units of measurement.

Publications that should be consulted for details of recommended metric practice and more complete information on the SI include the Standard Metric Practice Guide (designated E380-72) published by the American Society for Testing and Materials, 1916 Race St., Philadelphia, Pa. 19103 and also available from the American National Standards Institute, 1430 Broadway, New York, N. Y. 10018 as American National Standard Z210.1; ISO International Standard 1000, which covers rules for the use of SI units, multiples and submultiples, and is available from ANSI; and the National Bureau of Standards Special Publication 330, The International System of Units (SI).*

METRIC SYSTEMS OF MEASUREMENT

A metric system of measurement was first established in France in the years following the French Revolution, and various systems of metric units have been developed since that time. All metric unit systems are based, at least in part, on the International Metric Standards which are the metre and kilogram, or decimal multiples or sub-multiples of these standards.

In 1795, a metric system called the centimetre-gram-second (cgs) system was proposed, and was adopted in France in 1799. In 1873, the British Association for the Advancement of Science recommended the use of the cgs system, and since then it has been widely used in all branches of science throughout the world. From the base units in the cgs system are derived:

Unit of velocity = 1 centimetre per second.
Acceleration due to gravity (at Paris) = 981 centimetres per sec. per sec.
Unit of force = 1 dyne = 1/981 gram.
Unit of work = 1 erg = 1 dyne-centimetre.
Unit of power = 1 watt = 10,000,000 ergs per second.

Another metric system called the MKS (metre-kilogram-second) system of units was proposed by Professor G. Giorgi in 1902. In 1935, the International Electrotechnical Commission (IEC) accepted his recommendation that this system of units ot mechanics should be linked with the electro-magnetic units by the adoption of a fourth base unit. In 1950, the IEC adopted the ampere, the unit of electric current, as the fourth unit, and the MKSA system thus came into being.

A gravitational system of metric units, known as the technical system, is based on the metre, the kilogram as a force, and the second. It has been widely used in engineering. Because the standard of force is defined as the weight of the mass of the standard kilogram, the fundamental unit of force varies due to the difference in

* For sale by Superintendent of Documents, U. S. Government Printing Office, Washington, D. C. 20402.

gravitational pull at different locations around the earth. By international agreement, a standard value for acceleration due to gravity was chosen (9.81 meters per second squared) which for all practical measurements is approximately the same as the local value at the point of measurement.

The International System of Units (SI). — The Conference Generale des Poids et Mesures (CGPM) which is the body responsible for all international matters concerning the metric system, adopted in 1954, a rationalized and coherent system of units, based on the four MKSA units (see above), and including the ***kelvin*** as the unit of temperature and the *candela* as the unit of luminous intensity. In 1960, the CGPM formally named this system the Systeme International d'Unites, for which the abbreviation is SI in all languages. In 1971, the 14th CGPM adopted a seventh base unit, the *mole* which is the unit of quantity ("amount of substance").

In the period since the first metric system was established in France towards the end of the 18th century, most of the countries of the world have adopted a metric system. At the present time most of the industrially advanced metric-using countries are changing from their traditional metric system to SI. Those countries which are currently changing or considering change from the English system of measurement to metric, have the advantage that they can convert directly to the modernized system. The United Kingdom, which can be said to have led the now worldwide move to change from the English system, went straight to SI.

The SI, like the traditional metric system, is based on decimal arithmetic. For each physical quantity, units of different sizes are formed by multiplying or dividing a single base value by powers of 10. Thus, changes can be made very simply by adding zeros or shifting decimal points. For example, the meter is the basic unit of length; the kilometer is a multiple (1000 meters); and the millimeter is a submultiple (one-thousandth of a meter).

In the older metric systems, the simplicity of a series of units linked by powers of ten is an advantage for plain quantities such as length, but this simplicity is lost as soon as more complex units are encountered. For example, in different branches of science and engineering, energy may appear as the erg, the calorie, the kilogram meter, the liter atmosphere or the horsepower hour. In contrast, the SI provides only one basic unit for each physical quantity, and universality is thus achieved.

As mentioned above, there are seven base-units, which are for the basic quantities of length, mass, time, electric current, thermodynamic temperature, amount of substance and luminous intensity, expressed as the metre (m), the kilogram (kg), the second (s), the ampere (A), the kelvin (K), the mole (mol) and the candela (cd). The units are defined in the accompanying table.

The SI is a coherent system. A system is said to be coherent if the product or quotient of any two unit quantities in the system is the unit of the resultant quantity. For example, in a coherent system in which the foot is the unit of length, the square foot is the unit of area, whereas the acre is not.

Other physical quantities are derived from the base units. For example, the unit of velocity is the meter per second (m/s), which is a combination of the base units of length and time. The unit of acceleration is the meter per second squared (m/s^2). By applying Newton's second law of motion — force is proportional to mass multiplied by acceleration — the unit of force is obtained which is the kilogram metre per second squared ($kg \cdot m/s^2$). This unit is known as the newton or N. Work, or force times distance is the kilogram metre squared per second squared ($kg \cdot m^2/s^2$), which is the joule, (1 joule = 1 newton-metre), and energy is also expressed in these terms. The abbreviation for joule is J. Power or work per unit time is the kilogram metre squared per second cubed ($kg \cdot m^2/s^3$), which is the watt (1 watt = 1 joule per second = 1 newton-metre per second.) The abbreviation for watt is W. The term horsepower is not used in the SI and is replaced by the watt which together with multiples and sub-multiples — kilowatt and milliwatt, for example — is the same unit as that used in electrical work.

The use of the newton as the unit of force is of particular interest to engineers. In practical work using the English or traditional metric systems of measurements, it is a common practice to apply weight units as force units. Thus, the unit of force in those systems is that force which when applied to unit mass produces an acceleration g, rather than unit acceleration. The value of gravitational acceleration g varies around the earth, and thus the weight of a given mass also varies. In an effort to account for this minor error, the kilogram-force and pound-force were introduced, which are defined as the forces due to "standard gravity" acting on bodies of one kilogram or one pound mass respectively. The standard gravitational acceleration is taken as 9.80665 metres per second squared or 32.174 feet per second squared. The newton is defined as "that force which when applied to a body having a mass of one kilogram, gives it an acceleration of one metre per second squared." It is independent of g. As a result, the factor g disappears from a wide range of formulas in dynamics. However, in some formulas in statics, where the weight of a body is important rather than its mass, g does appear where it was formerly absent (the weight of a mass of W kilograms is equal to a force of Wg newtons, where g = approximately 9.81 metre per second squared). Details concerning the use of SI units in mechanics calculations are given in the Mechanics section of *Machinery's Handbook* and the use of SI units in strength of materials calculations is covered in the section on that subject.

Decimal multiples and sub-multiples of the SI units are formed by means of the prefixes given in the following table, which represent the numerical factors shown.

Factors and Prefixes for Forming Decimal Multiples and Sub-multiples of the SI Units

Factor by which the unit is multiplied	Prefix	Symbol	Factor by which the unit is multiplied	Prefix	Symbol
10^{12}	tera	T	10^{-2}	centi	c
10^{9}	giga	G	10^{-3}	milli	m
10^{6}	mega	M	10^{-6}	micro	μ
10^{3}	kilo	k	10^{-9}	nano	n
10^{2}	hecto	h	10^{-12}	pico	p
10	deka	da	10^{-15}	femto	f
10^{-1}	deci	d	10^{-18}	atto	a

International System (SI) Units

PHYSICAL QUANTITY	NAME OF UNIT	UNIT SYMBOL	DEFINITION
		Basic SI Units	
Length	metre	m	1,650,763.73 wavelengths in vacuo of the radiation corresponding to the transition between the energy levels $2p_{10}$ and $5d_5$ of the krypton — 86 atom.
Mass	kilogram	kg	Mass of the international prototype which is in the custody of the Bureau International des Poids et Mesures (BIPM) at Sèvres, near Paris.
Time	second	s	The duration of 9,192,631,770 periods of the radiation corresponding to the transition between the two hyperfine levels of the ground state of the cesium-133 atom.
Electric Current	ampere	A	The constant current which, if maintained in two parallel rectilinear conductors of infinite length, of negligible circular cross section, and placed at a distance of one metre apart in a vacuum, would produce between these conductors a force equal to 2×10^{-7} N/m length.
Thermodynamic Temperature	degree Kelvin	K	The fraction 1/273.16 of the thermodynamic temperature of the triple point of water.
Amount of Substance	mole	mol	The amount of substance of a system which contains as many elementary entities as there are atoms in 0.012 kilogram of carbon 12.
Luminous Intensity	candela	cd	The luminous intensity, in the perpendicular direction, of a surface of 1/600,000 square metre of a black body at the temperature of freezing platinum under a pressure of 101,325 newtons per square metre.
		SI Units Having Special Names	
Force	newton	$N = kg{\cdot}m/s^2$	That force which, when applied to a body having a mass of one kilogramme, gives it an acceleration of one metre per second squared.
Work, Energy, Quantity of Heat	joule	$J = N{\cdot}m$	The work done when the point of application of a force of one newton is displaced through a distance of one metre in the direction of the force.
Power	watt	$W = J/s$	One joule per second.
Electric Charge	coulomb	$C = A{\cdot}s$	The quantity of electricity transported in one second by a current of one ampere.
Electric Potential	volt	$V = W/A$	The difference of potential between two points of a conducting wire carrying a constant current of one ampere, when the power dissipated between these points is equal to one watt.
Electric Capacitance	farad	$F = C/V$	The capacitance of a capacitor between the plates of which there appears a difference of potential of one volt when it is charged by a quantity of electricity equal to one coulomb.

International System (SI) Units (*Continued*)

PHYSICAL QUANTITY	NAME OF UNIT	UNIT SYMBOL	DEFINITION
SI Units Having Special Names			
Electric Resistance	ohm	$\Omega = V/A$	The resistance between two points of a conductor when a constant difference of potential of one volt, applied between these two points, produces in this conductor a current of one ampere, this conductor not being the source of any electromotive force.
Magnetic Flux	weber	$Wb = V{\cdot}s$	The flux which, linking a circuit of one turn produces in it an electromotive force of one volt as it is reduced to zero at a uniform rate in one second.
Inductance	henry	**H = Wb/A**	The inductance of a closed circuit in which an electromotive force of one volt is produced when the electric current in the circuit varies uniformly at the rate of one ampere per second.
Luminous Flux	lumen	$lm = cd{\cdot}sr$	The flux emitted within a unit solid angle of one steradian by a point source having a uniform intensity of one candela.
Illumination	lux	$lx = lm/m^2$	An illumination of one lumen per square metre.

International System (SI) Units with Complex Names

PHYSICAL QUANTITY	SI UNIT	UNIT SYMBOL
SI Units Having Complex Names		
Area	square metre	m^2
Volume	cubic metre	m^3
Frequency	**hertz***	**Hz**
Density (Mass Density)	kilogram per cubic metre	kg/m^3
Velocity	metre per second	m/s
Angular Velocity	radian per second	rad/s
Acceleration	metre per second squared	m/s^2
Angular Acceleration	radian per second squared	rad/s^2
Pressure	newton per square metre	N/m^2
Surface Tension	newton per metre	N/m
Dynamic Viscosity	newton second per metre squared	$N\ s/m^2$
Kinematic Viscosity, Diffusion Coefficient	metre squared per second	m^2/s
Thermal Conductivity	watt per metre degree Kelvin	W/(m °K)
Electric Field Strength	volt per metre	V/m
Magnetic Flux Density	**tesla†**	**T**
Magnetic Field Strength	ampere per metre	A/m
Luminance	candela per square metre	cd/m^2

*** Hz = cycle/second.**

† T = weber/metre².

Metric Conversion Factors

(Symbols of SI units, multiples and submultiples are given in parentheses in the right-hand column)

Multiply	By	To Obtain
	LENGTH	
centimetre	0.03280840	foot
centimetre	0.3937008	inch
fathom	1.8288*	metre (m)
foot	0.3048*	metre (m)
foot	30.48*	centimetre (cm)
foot	304.8*	millimetre (mm)
inch	0.0254*	metre (m)
inch	2.54*	centimetre (cm)
inch	25.4*	millimetre (mm)
kilometre	0.6213712	mile [U. S. statute]
metre	39.37008	inch
metre	0.5468066	fathom
metre	3.280840	foot
metre	0.1988388	rod
metre	1.093613	yard
metre	0.0006213712	mile [U. S. statute]
microinch	0.0254*	micrometre [micron] (μm)
micrometre [micron]	39.37008	microinch
mile [U. S. statute]	1609.344*	metre (m)
mile [U. S. statute]	1.609344*	kilometre (km)
millimetre	0.003280840	foot
millimetre	0.03937008	inch
rod	5.0292*	metre (m)
yard	0.9144*	metre (m)
	AREA	
acre	4046.856	metre^2 (m^2)
acre	0.4046856	hectare
centimetre^2	0.1550003	inch^2
centimetre^2	0.001076391	foot^2
foot^2	0.09290304*	metre^2 (m^2)
foot^2	929.0304*	centimetre^2 (cm^2)
foot^2	92,903.04*	millimetre^2 (mm^2)
hectare	2.471054	acre
inch^2	645.16*	millimetre^2 (mm^2)
inch^2	6.4516*	centimetre^2 (cm^2)
inch^2	0.00064516*	metre^2 (m^2)
metre^2	1550.003	inch^2
metre^2	10.763910	foot^2
metre^2	1.195990	yard^2
metre^2	0.0002471054	acre
millimetre^2	0.0001076387	foot^2
millimetre^2	0.001550003	inch^2
yard^2	0.8361274	metre^2 (m^2)

* Where an asterisk is shown, the figure is exact.

Metric Conversion Factors (*Continued*)

Multiply	By	To Obtain
VOLUME (including CAPACITY)		
centimetre^3	0.06102376	inch^3
foot^3	0.02831685	metre^3 (m^3)
foot^3	28.31685	litre
gallon [U. K. liquid]	0.004546092	metre^3 (m^3)
gallon [U. K. liquid]	4.546092	litre
gallon [U. S. liquid]	0.003785412	metre^3 (m^3)
gallon [U. S. liquid]	3.785412	litre
inch^3	16,387.06	millimetre^3 (mm^3)
inch^3	16.38706	centimetre^3 (cm^3)
inch^3	0.00001638706	metre^3 (m^3)
litre	0.001*	metre^3 (m^3)
litre	0.2199692	gallon [U. K. liquid]
litre	0.2641720	gallon [U. S. liquid]
litre	0.03531466	foot^3
metre^3	219.9692	gallon [U. K. liquid]
metre^3	264.1720	gallon [U. S. liquid]
metre^3	35.31466	foot^3
metre^3	1.307951	yard^3
metre^3	1000.*	litre
metre^3	61,023.76	inch^3
millimetre^3	0.00006102376	inch^3
yard^3	0.7645549	metre^3 (m^3)
VELOCITY, ACCELERATION, and FLOW		
centimetre/second	1.968504	foot/minute
centimetre/second	0.03280840	foot/second
centimetre/minute	0.3937008	inch/minute
foot/hour	0.00008466667	metre/second (m/s)
foot/hour	0.00508*	metre/minute
foot/hour	0.3048*	metre/hour
foot/minute	0.508*	centimetre/second
foot/minute	18.288*	metre/hour
foot/minute	0.3048*	metre/minute
foot/minute	0.00508*	metre/second (m/s)
foot/second	30.48*	centimetre/second
foot/second	18.288*	metre/minute
foot/second	0.3048*	metre/second (m/s)
foot/second^2	0.3048*	metre/second^2 (m/s^2)
foot^3/minute	28.31685	litre/minute
foot^3/minute	0.0004719474	metre^3/second (m^3/s)
gallon [U. S. liquid]/min.	0.003785412	metre^3/minute
gallon [U. S. liquid]/min.	0.00006309020	metre^3/second (m^3/s)
gallon [U. S. liquid]/min.	0.06309020	litre/second
gallon [U. S. liquid]/min.	3.785412	litre/minute
gallon [U. K. liquid]/min.	0.004546092	metre^3/minute
gallon [U. K. liquid]/min.	0.00007576820	metre^3/second (m^3/s)
inch/minute	25.4*	millimetre/minute
inch/minute	2.54*	centimetre/minute
inch/minute	0.0254*	metre/minute
inch/second^2	0.0254*	metre/second^2 (m/s^2)

* Where an asterisk is shown, the figure is exact.

Metric Conversion Factors (*Continued*)

Multiply	By	To Obtain
VELOCITY, ACCELERATION, and FLOW (*Continued*)		
kilometre/hour	0.6213712	mile/hour [U. S. statute]
litre/minute	0.03531466	$foot^3$/minute
litre/minute	0.2641720	gallon [U. S. liquid]/minute
litre/second	15.85032	gallon [U. S. liquid]/minute
mile/hour	1.609344*	kilometre/hour
millimetre/minute	0.03937008	inch/minute
metre/second	11,811.02	foot/hour
metre/second	196.8504	foot/minute
metre/second	3.280840	foot/second
$metre/second^2$	3.280840	$foot/second^2$
$metre/second^2$	39.37008	$inch/second^2$
metre/minute	3.280840	foot/minute
metre/minute	0.05468067	foot/second
metre/minute	39.37008	inch/minute
metre/hour	3.280840	foot/hour
metre/hour	0.05468067	foot/minute
$metre^3$/second	2118.880	$foot^3$/minute
$metre^3$/second	13.198.15	gallon [U. K. liquid]/minute
$metre^3$/second	15.850.32	gallon [U. S. liquid]/minute
$metre^3$/minute	219.9692	gallon [U. K. liquid]/minute
$metre^3$/minute	264.1720	gallon [U. S. liquid]/minute
MASS and DENSITY		
grain [1/7000 lb avoirdupois]	0.06479891	gram (g)
gram	15.43236	grain
gram	0.001*	kilogram (kg)
gram	0.03527397	ounce [avoirdupois]
gram	0.03215074	ounce [troy]
$gram/centimetre^3$	0.03612730	$pound/inch^3$
hundredweight [long]	50.80235	kilogram (kg)
hundredweight [short]	45.35924	kilogram (kg)
kilogram	1000.*	gram (g)
kilogram	35.27397	ounce [avoirdupois]
kilogram	32.15074	ounce [troy]
kilogram	2.204622	pound [avoirdupois]
kilogram	0.06852178	slug
kilogram	0.0009842064	ton [long]
kilogram	0.001102311	ton [short]
kilogram	0.001*	ton [metric]
kilogram	0.001*	tonne
kilogram	0.01968413	hundredweight [long]
kilogram	0.02204622	hundredweight [short]
$kilogram/metre^3$	0.06242797	$pound/foot^3$
$kilogram/metre^3$	0.008345406	pound/gallon [U. K. liquid]
$kilogram/metre^3$	0.01002242	pound/gallon [U. S. liquid]
ounce [avoirdupois]	28.34952	gram (g)
ounce [avoirdupois]	0.02834952	kilogram (kg)

* Where an asterisk is shown, the figure is exact.

Metric Conversion Factors (*Continued*)

Multiply	By	To Obtain
MASS and DENSITY (*Continued*)		
ounce [troy]	31.10348	gram (g)
ounce [troy]	0.03110348	kilogram (kg)
pound [avoirdupois]	0.4535924	kilogram (kg)
pound/foot³	16.01846	kilogram/metre³ (kg/m³)
pound/inch³	27.67990	gram/centimetre³ (g/cm³)
pound/gal [U. S. liquid]	119.8264	kilogram/metre³ (kg/m³)
pound/gal [U. K. liquid]	99.77633	kilogram/metre³ (kg/m³)
slug	14.59390	kilogram (kg)
ton [long 2240 lb]	1016.047	kilogram (kg)
ton [short 2000 lb]	907.1847	kilogram (kg)
ton [metric]	1000.*	kilogram (kg)
tonne	1000.*	kilogram (kg)
FORCE and FORCE/LENGTH		
dyne	0.00001*	newton (N)
kilogram-force	9.806650*	newton (N)
kilopond	9.806650*	newton (N)
newton	0.1019716	kilogram-force
newton	0.1019716	kilopond
newton	0.2248089	pound-force
newton	100,000.*	dyne
newton	7.23301	poundal
newton	3.596942	ounce-force
newton/metre	0.005710148	pound/inch
newton/metre	0.06852178	pound/foot
ounce-force	0.2780139	newton (N)
pound-force	4.448222	newton (N)
poundal	0.1382550	newton (N)
pound/inch	175.1268	newton/metre (N/m)
pound/foot	14.59390	newton/metre (N/m)
BENDING MOMENT or TORQUE		
dyne-centimetre	0.0000001*	newton-metre (N · m)
kilogram-metre	9.806650*	newton-metre (N · m)
ounce-inch	7.061552	newton-millimetre
ounce-inch	0.007061552	newton-metre (N · m)
newton-metre	0.7375621	pound-foot
newton-metre	10,000,000.*	dyne-centimetre
newton-metre	0.1019716	kilogram-metre
newton-metre	141.6119	ounce-inch
newton-millimetre	0.1416119	ounce-inch
pound-foot	1.355818	newton-metre (N · m)

* Where an asterisk is shown, the figure is exact.

Metric Conversion Factors (*Continued*)

Multiply	By	To Obtain
MOMENT OF INERTIA and SECTION MODULUS		
moment of inertia [kg · m^2]	23.73036	pound-foot^2
moment of inertia [kg · m^2]	3417.171	pound-inch^2
moment of inertia [lb · ft^2]	0.04214011	kilogram-metre^2 (kg · m^2)
moment of inertia [lb · inch^2]	0.0002926397	kilogram-metre^2 (kg · m^2)
moment of section [foot^4]	0.008630975	metre^4 (m^4)
moment of section [inch^4]	41.62314	centimetre^4
moment of section [metre^4]	115.8618	foot^4
moment of section [centimetre^4]	0.02402510	inch^4
section modulus [foot^3]	0.02831685	metre^3 (m^3)
section modulus [inch^3]	0.00001638706	metre^3 (m^3)
section modulus [metre^3]	35.31466	foot^3
section modulus [metre^3]	61,023.76	inch^3
MOMENTUM		
kilogram-metre/second	7.233011	pound-foot/second
kilogram-metre/second	86.79614	pound-inch/second
pound-foot/second	0.1382550	kilogram-metre/second (kg · m/s)
pound-inch/second	0.01152125	kilogram-metre/second (kg · m/s)
PRESSURE and STRESS		
atmosphere [14.6959 lb/inch^2]	101,325.	pascal (Pa)
bar	100,000.*	pascal (Pa)
bar	14.50377	pound/inch^2
bar	100,000.*	newton/metre^2 (N/m^2)
hectobar	0.6474898	ton [long]/inch^2
kilogram/centimetre^2	14.22334	pound/inch^2
kilogram/metre^2	9.806650*	newton/metre^2 (N/m^2)
kilogram/metre^2	9.806650*	pascal (Pa)
kilogram/metre^2	0.2048161	pound/foot^2
kilonewton/metre^2	0.1450377	pound/inch^2
newton/centimetre^2	1.450377	pound/inch^2
newton/metre^2	0.00001*	bar
newton/metre^2	1.0*	pascal (Pa)
newton/metre^2	0.0001450377	pound/inch^2
newton/metre^2	0.1019716	kilogram/metre^2
newton/millimetre^2	145.0377	pound/inch^2
pascal	0.00000986923	atmosphere
pascal	0.00001*	bar
pascal	0.1019716	kilogram/metre^2
pascal	1.0*	newton/metre^2 (N/m^2)
pascal	0.02088543	pound/foot^2
pascal	0.0001450377	pound/inch^2

* Where an asterisk is shown, the figure is exact.

Metric Conversion Factors (*Continued*)

Multiply	By	To Obtain
	PRESSURE and STRESS (*Continued*)	
pound/foot^2	4.882429	kilogram/metre^2
pound/foot^2	47.88026	pascal (Pa)
pound/inch^2	0.06894757	bar
pound/inch^2	0.07030697	kilogram/centimetre^2
pound/inch^2	0.6894757	newton/centimetre^2
pound/inch^2	6.894757	kilonewton/metre^2
pound/inch^2	6894.757	newton/metre^2 (N/m^2)
pound/inch^2	0.006894757	newton/millimetre^2 (N/mm^2)
pound/inch^2	6894.757	pascal (Pa)
ton [long]/inch^2	1.544426	hectobar
	ENERGY and WORK	
Btu [International Table]	1055.056	joule (J)
Btu [mean]	1055.87	joule (J)
calorie [mean]	4.19002	joule (J)
foot-pound	1.355818	joule (J)
foot-poundal	0.04214011	joule (J)
joule	0.0009478170	Btu [International Table]
joule	0.0009470863	Btu [mean]
joule	0.2386623	calorie [mean]
joule	0.7375621	foot-pound
joule	23.73036	foot-poundal
joule	0.9998180	joule [International U. S.]
joule	0.9999830	joule [U. S. legal, 1948]
joule [International U. S.]	1.000182	joule (J)
joule [U. S. legal, 1948]	1.000017	joule (J)
joule	.0002777778	watt-hour
watt-hour	3600.*	joule (J)
	POWER	
Btu [International Table]/hour	0.2930711	watt (W)
foot-pound/hour	0.0003766161	watt (W)
foot-pound/minute	0.02259697	watt (W)
horsepower [550 ft-lb/s]	0.7456999	kilowatt (kW)
horsepower [550 ft-lb/s]	745.6999	watt (W)
horsepower [electric]	746.*	watt (W)
horsepower [metric]	735.499	watt (W)
horsepower [U. K.]	745.70	watt (W)
kilowatt	1.341022	horsepower [550 ft-lb/s]
watt	2655.224	foot-pound/hour
watt	44.25372	foot-pound/minute
watt	0.001341022	horsepower [550 ft-lb/s]
watt	0.001340483	horsepower [electric]
watt	0.001359621	horsepower [metric]
watt	0.001341022	horsepower [U. K.]
watt	3.412141	Btu [International Table]/hour

* Where an asterisk is shown, the figure is exact.

Metric Conversion Factors (*Continued*)

Multiply	By	To Obtain
VISCOSITY		
centipoise	0.001*	pascal-second (Pa · s)
centistoke	0.000001*	metre²/second (m²/s)
metre²/second	1,000,000.*	centistoke
metre²/second	10,000.*	stoke
pascal-second	1000.*	centipoise
pascal-second	10.*	poise
poise	0.1*	pascal-second (Pa · s)
stoke	0.0001*	metre²/second (m²/s)
TEMPERATURE		
To Convert From	To	Use Formula
temperature Celsius, t_C	temperature Kelvin, t_K	$t_K = t_C + 273.15$
temperature Fahrenheit, t_F	temperature Kelvin, t_K	$t_K = (t_F + 459.67)/1.8$
temperature Celsius, t_C	temperature Fahrenheit, t_F	$t_F = 1.8\, t_C + 32$
temperature Fahrenheit, t_F	temperature Celsius, t_C	$t_C = (t_F - 32)/1.8$
temperature Kelvin, t_K	temperature Celsius, t_C	$t_C = t_K - 273.15$
temperature Kelvin, t_K	temperature Fahrenheit, t_F	$t_F = 1.8\, t_K - 459.67$
temperature Kelvin, t_K	temperature Rankine, t_R	$t_R = 9/5\, t_K$
temperature Rankine, t_R	temperature Kelvin, t_K	$t_K = 5/9\, t_R$

* Where an asterisk is shown, the figure is exact.

Use of Conversion Tables. — On this and following pages tables are given which permit conversion from English to metric units and vice versa over a wide range of values. Where the desired value cannot be obtained directly from these tables, a simple addition of two or more values taken directly from the table will suffice as shown in the following examples:

Example 1: Find the millimetre equivalent of 0.4476 inches.

.4 in. = 10.16000 mm
.04 in. = 1.01600 mm
.007 in. = .17780 mm
.0006 in. = .01524 mm

.4476 in. = 11.36904 mm

Example 2: Find the inch equivalent of 84.9 mm.

80. mm = 3.14961 in.
4. mm = 0.15748 in.
0.9 mm = 0.03543 in.

84.9 mm = 3.34252 in.

Inch—Millimetre and Inch—Centimetre Conversion Table*
(Based on 1 inch = 25.4 millimetres, exactly)

INCHES TO MILLIMETRES

in.	mm	in.	mm	in.	mm	in.	mm	in.	mm	in.	mm
10	254.00000	1	25.40000	.1	2.54000	.01	.25400	.001	.02540	.0001	.00254
20	508.00000	2	50.80000	.2	5.08000	.02	.50800	.002	.05080	.0002	.00508
30	762.00000	3	76.20000	.3	7.62000	.03	.76200	.003	.07620	.0003	.00762
40	1,016.00000	4	101.60000	.4	10.16000	.04	1.01600	.004	.10160	.0004	.01016
50	1,270.00000	5	127.00000	.5	12.70000	.05	1.27000	.005	.12700	.0005	.01270
60	1,524.00000	6	152.40000	.6	15.24000	.06	1.52400	.006	.15240	.0006	.01524
70	1,778.00000	7	177.80000	.7	17.78000	.07	1.77800	.007	.17780	.0007	.01778
80	2,032.00000	8	203.20000	.8	20.32000	.08	2.03200	.008	.20320	.0008	.02032
90	2,286.00000	9	228.60000	.9	22.86000	.09	2.28600	.009	.22860	.0009	.02286
100	2,540.00000	10	254.00000	1.0	25.40000	.10	2.54000	.010	.25400	.0010	.02540

MILLIMETRES TO INCHES

mm	in.	mm	in.	mm.	in.	mm	in.	mm	in.	mm	in.
100	3.93701	10	.39370	1	.03937	.1	.00394	.01	.00039	.001	.00004
200	7.87402	20	.78740	2	.07874	.2	.00787	.02	.00079	.002	.00008
300	11.81102	30	1.18110	3	.11811	.3	.01181	.03	.00118	.003	.00012
400	15.74803	40	1.57480	4	.15748	.4	.01575	.04	.00157	.004	.00016
500	19.68504	50	1.96850	5	.19685	.5	.01969	.05	.00197	.005	.00020
600	23.62205	60	2.36220	6	.23622	.6	.02362	.06	.00236	.006	.00024
700	27.55906	70	2.75591	7	.27559	.7	.02756	.07	.00276	.007	.00028
800	31.49606	80	3.14961	8	.31496	.8	.03150	.08	.00315	.008	.00031
900	35.43307	90	3.54331	9	.35433	.9	.03543	.09	.00354	.009	.00035
1,000	39.37008	100	3.93701	10	.39370	1.0	.03937	.10	.00394	.010	.00039

* For inches to centimetres, shift decimal point in mm column one place to left and read centimetres, thus:

40 in. = 1016 mm = 101.6 cm

For centimetres to inches, shift decimal point of centimetre value one place to right and enter mm column, thus:

70 cm = 700 mm = 27.55906 inches

Decimals of an Inch to Millimetres
(Based on 1 inch = 25.4 millimetres, exactly)

Inches	0.000	0.001	0.002	0.003	0.004	0.005	0.006	0.007	0.008	0.009
	Millimetres									
0.000	...	0.0254	0.0508	0.0762	0.1016	0.1270	0.1524	0.1778	0.2032	0.2286
0.010	0.2540	0.2794	0.3048	0.3302	0.3556	0.3810	0.4064	0.4318	0.4572	0.4826
0.020	0.5080	0.5334	0.5588	0.5842	0.6096	0.6350	0.6604	0.6858	0.7112	0.7366
0.030	0.7620	0.7874	0.8128	0.8382	0.8636	0.8890	0.9144	0.9398	0.9652	0.9906
0.040	1.0160	1.0414	1.0668	1.0922	1.1176	1.1430	1.1684	1.1938	1.2192	1.2446
0.050	1.2700	1.2954	1.3208	1.3462	1.3716	1.3970	1.4224	1.4478	1.4732	1.4986
0.060	1.5240	1.5494	1.5748	1.6002	1.6256	1.6510	1.6764	1.7018	1.7272	1.7526
0.070	1.7780	1.8034	1.8288	1.8542	1.8796	1.9050	1.9304	1.9558	1.9812	2.0066
0.080	2.0320	2.0574	2.0828	2.1082	2.1336	2.1590	2.1844	2.2098	2.2352	2.2606
0.090	2.2860	2.3114	2.3368	2.3622	2.3876	2.4130	2.4384	2.4638	2.4892	2.5146
0.100	2.5400	2.5654	2.5908	2.6162	2.6416	2.6670	2.6924	2.7178	2.7432	2.7686
0.110	2.7940	2.8194	2.8448	2.8702	2.8956	2.9210	2.9464	2.9718	2.9972	3.0226
0.120	3.0480	3.0734	3.0988	3.1242	3.1496	3.1750	3.2004	3.2258	3.2512	3.2766
0.130	3.3020	3.3274	3.3528	3.3782	3.4036	3.4290	3.4544	3.4798	3.5052	3.5306
0.140	3.5560	3.5814	3.6068	3.6322	3.6576	3.6830	3.7084	3.7338	3.7592	3.7846
0.150	3.8100	3.8354	3.8608	3.8862	3.9116	3.9370	3.9624	3.9878	4.0132	4.0386
0.160	4.0640	4.0894	4.1148	4.1402	4.1656	4.1910	4.2164	4.2418	4.2672	4.2926
0.170	4.3180	4.3434	4.3688	4.3942	4.4196	4.4450	4.4704	4.4958	4.5212	4.5466
0.180	4.5720	4.5974	4.6228	4.6482	4.6736	4.6990	4.7244	4.7498	4.7752	4.8006
0.190	4.8260	4.8514	4.8768	4.9022	4.9276	4.9530	4.9784	5.0038	5.0292	5.0546
0.200	5.0800	5.1054	5.1308	5.1562	5.1816	5.2070	5.2324	5.2578	5.2832	5.3086
0.210	5.3340	5.3594	5.3848	5.4102	5.4356	5.4610	5.4864	5.5118	5.5372	5.5626
0.220	5.5880	5.6134	5.6388	5.6642	5.6896	5.7150	5.7404	5.7658	5.7912	5.8166
0.230	5.8420	5.8674	5.8928	5.9182	5.9436	5.9690	5.9944	6.0198	6.0452	6.0706
0.240	6.0960	6.1214	6.1468	6.1722	6.1976	6.2230	6.2484	6.2738	6.2992	6.3246
0.250	6.3500	6.3754	6.4008	6.4262	6.4516	6.4770	6.5024	6.5278	6.5532	6.5786
0.260	6.6040	6.6294	6.6548	6.6802	6.7056	6.7310	6.7564	6.7818	6.8072	6.8326
0.270	6.8580	6.8834	6.9088	6.9342	6.9596	6.9850	7.0104	7.0358	7.0612	7.0866
0.280	7.1120	7.1374	7.1628	7.1882	7.2136	7.2390	7.2644	7.2898	7.3152	7.3406
0.290	7.3660	7.3914	7.4168	7.4422	7.4676	7.4930	7.5184	7.5438	7.5692	7.5946
0.300	7.6200	7.6454	7.6708	7.6962	7.7216	7.7470	7.7724	7.7978	7.8232	7.8486
0.310	7.8740	7.8994	7.9248	7.9502	7.9756	8.0010	8.0264	8.0518	8.0772	8.1026
0.320	8.1280	8.1534	8.1788	8.2042	8.2296	8.2550	8.2804	8.3058	8.3312	8.3566
0.330	8.3820	8.4074	8.4328	8.4582	8.4836	8.5090	8.5344	8.5598	8.5852	8.6106
0.340	8.6360	8.6614	8.6868	8.7122	8.7376	8.7630	8.7884	8.8138	8.8392	8.8646
0.350	8.8900	8.9154	8.9408	8.9662	8.9916	9.0170	9.0424	9.0678	9.0932	9.1186
0.360	9.1440	9.1694	9.1948	9.2202	9.2456	9.2710	9.2964	9.3218	9.3472	9.3726
0.370	9.3980	9.4234	9.4488	9.4742	9.4996	9.5250	9.5504	9.5758	9.6012	9.6266
0.380	9.6520	9.6774	9.7028	9.7282	9.7536	9.7790	9.8044	9.8298	9.8552	9.8806
0.390	9.9060	9.9314	9.9568	9.9822	10.0076	10.0330	10.0584	10.0838	10.1092	10.1346
0.400	10.1600	10.1854	10.2108	10.2362	10.2616	10.2870	10.3124	10.3378	10.3632	10.3886
0.410	10.4140	10.4394	10.4648	10.4902	10.5156	10.5410	10.5664	10.5918	10.6172	10.6426
0.420	10.6680	10.6934	10.7188	10.7442	10.7696	10.7950	10.8204	10.8458	10.8712	10.8966
0.430	10.9220	10.9474	10.9728	10.9982	11.0236	11.0490	11.0744	11.0998	11.1252	11.1506
0.440	11.1760	11.2014	11.2268	11.2522	11.2776	11.3030	11.3284	11.3538	11.3792	11.4046
0.450	11.4300	11.4554	11.4808	11.5062	11.5316	11.5570	11.5824	11.6078	11.6332	11.6586
0.460	11.6840	11.7094	11.7348	11.7602	11.7856	11.8110	11.8364	11.8618	11.8872	11.9126
0.470	11.9380	11.9634	11.9888	12.0142	12.0396	12.0650	12.0904	12.1158	12.1412	12.1666
0.480	12.1920	12.2174	12.2428	12.2682	12.2936	12.3190	12.3444	12.3698	12.3952	12.4206
0.490	12.4460	12.4714	12.4968	12.5222	12.5476	12.5730	12.5984	12.6238	12.6492	12.6746

Use previous table to obtain whole inch equivalents to add to decimal equivalents above. All values given in this table are exact; figures to the right of the last place figures are all zeros.

Decimals of an Inch to Millimetres
(Based on 1 inch = 25.4 millimetres, exactly)

Inches	0.000	0.001	0.002	0.003	0.004	0.005	0.006	0.007	0.008	0.009
	Millimetres									
0.500	12.7000	12.7254	12.7508	12.7762	12.8016	12.8270	12.8524	12.8778	12.9032	12.9286
0.510	12.9540	12.9794	13.0048	13.0302	13.0556	13.0810	13.1064	13.1318	13.1572	13.1826
0.520	13.2080	13.2334	13.2588	13.2842	13.3096	13.3350	13.3604	13.3858	13.4112	13.4366
0.530	13.4620	13.4874	13.5128	13.5382	13.5636	13.5890	13.6144	13.6398	13.6652	13.6906
0.540	13.7160	13.7414	13.7668	13.7922	13.8176	13.8430	13.8684	13.8938	13.9192	13.9446
0.550	13.9700	13.9954	14.0208	14.0462	14.0716	14.0970	14.1224	14.1478	14.1732	14.1986
0.560	14.2240	14.2494	14.2748	14.3002	14.3256	14.3510	14.3764	14.4018	14.4272	14.4526
0.570	14.4780	14.5034	14.5288	14.5542	14.5796	14.6050	14.6304	14.6558	14.6812	14.7066
0.580	14.7320	14.7574	14.7828	14.8082	14.8336	14.8590	14.8844	14.9098	14.9352	14.9606
0.590	14.9860	15.0114	15.0368	15.0622	15.0876	15.1130	15.1384	15.1638	15.1892	15.2146
0.600	15.2400	15.2654	15.2908	15.3162	15.3416	15.3670	15.3924	15.4178	15.4432	15.4686
0.610	15.4940	15.5194	15.5448	15.5702	15.5956	15.6210	15.6464	15.6718	15.6972	15.7226
0.620	15.7480	15.7734	15.7988	15.8242	15.8496	15.8750	15.9004	15.9258	15.9512	15.9766
0.630	16.0020	16.0274	16.0528	16.0782	16.1036	16.1290	16.1544	16.1798	16.2052	16.2306
0.640	16.2560	16.2814	16.3068	16.3322	16.3576	16.3830	16.4084	16.4338	16.4592	16.4846
0.650	16.5100	16.5354	16.5608	16.5862	16.6116	16.6370	16.6624	16.6878	16.7132	16.7386
0.660	16.7640	16.7894	16.8148	16.8402	16.8656	16.8910	16.9164	16.9418	16.9672	16.9926
0.670	17.0180	17.0434	17.0688	17.0942	17.1196	17.1450	17.1704	17.1958	17.2212	17.2466
0.680	17.2720	17.2974	17.3228	17.3482	17.3736	17.3990	17.4244	17.4498	17.4752	17.5006
0.690	17.5260	17.5514	17.5768	17.6022	17.6276	17.6530	17.6784	17.7038	17.7292	17.7546
0.700	17.7800	17.8054	17.8308	17.8562	17.8816	17.9070	17.9324	17.9578	17.9832	18.0086
0.710	18.0340	18.0594	18.0848	18.1102	18.1356	18.1610	18.1864	18.2118	18.2372	18.2626
0.720	18.2880	18.3134	18.3388	18.3642	18.3896	18.4150	18.4404	18.4658	18.4912	18.5166
0.730	18.5420	18.5674	18.5928	18.6182	18.6436	18.6690	18.6944	18.7198	18.7452	18.7706
0.740	18.7960	18.8214	18.8468	18.8722	18.8976	18.9230	18.9484	18.9738	18.9992	19.0246
0.750	19.0500	19.0754	19.1008	19.1262	19.1516	19.1770	19.2024	19.2278	19.2532	19.2786
0.760	19.3040	19.3294	19.3548	19.3802	19.4056	19.4310	19.4564	19.4818	19.5072	19.5326
0.770	19.5580	19.5834	19.6088	19.6342	19.6596	19.6850	19.7104	19.7358	19.7612	19.7866
0.780	19.8120	19.8374	19.8628	19.8882	19.9136	19.9390	19.9644	19.9898	20.0152	20.0406
0.790	20.0660	20.0914	20.1168	20.1422	20.1676	20.1930	20.2184	20.2438	20.2692	20.2946
0.800	20.3200	20.3454	20.3708	20.3962	20.4216	20.4470	20.4724	20.4978	20.5232	20.5486
0.810	20.5740	20.5994	20.6248	20.6502	20.6756	20.7010	20.7264	20.7518	20.7772	20.8026
0.820	20.8280	20.8534	20.8788	20.9042	20.9296	20.9550	20.9804	21.0058	21.0312	21.0566
0.830	21.0820	21.1074	21.1328	21.1582	21.1836	21.2090	21.2344	21.2598	21.2852	21.3106
0.840	21.3360	21.3614	21.3868	21.4122	21.4376	21.4630	21.4884	21.5138	21.5392	21.5646
0.850	21.5900	21.6154	21.6408	21.6662	21.6916	21.7170	21.7424	21.7678	21.7932	21.8186
0.860	21.8440	21.8694	21.8948	21.9202	21.9456	21.9710	21.9964	22.0218	22.0472	22.0726
0.870	22.0980	22.1234	22.1488	22.1742	22.1996	22.2250	22.2504	22.2758	22.3012	22.3266
0.880	22.3520	22.3774	22.4028	22.4282	22.4536	22.4790	22.5044	22.5298	22.5552	22.5806
0.890	22.6060	22.6314	22.6568	22.6822	22.7076	22.7330	22.7584	22.7838	22.8092	22.8346
0.900	22.8600	22.8854	22.9108	22.9362	22.9616	22.9870	23.0124	23.0378	23.0632	23.0886
0.910	23.1140	23.1394	23.1648	23.1902	23.2156	23.2410	23.2664	23.2918	23.3172	23.3426
0.920	23.3680	23.3934	23.4188	23.4442	23.4696	23.4950	23.5204	23.5458	23.5712	23.5966
0.930	23.6220	23.6474	23.6728	23.6982	23.7236	23.7490	23.7744	23.7998	23.8252	23.8506
0.940	23.8760	23.9014	23.9268	23.9522	23.9776	24.0030	24.0284	24.0538	24.0792	24.1046
0.950	24.1300	24.1554	24.1808	24.2062	24.2316	24.2570	24.2824	24.3078	24.3332	24.3586
0.960	24.3840	24.4094	24.4348	24.4602	24.4856	24.5110	24.5364	24.5618	24.5872	24.6126
0.970	24.6380	24.6634	24.6888	24.7142	24.7396	24.7650	24.7904	24.8158	24.8412	24.8666
0.980	24.8920	24.9174	24.9428	24.9682	24.9936	25.0190	25.0444	25.0698	25.0952	25.1206
0.990	25.1460	25.1714	25.1968	25.2222	25.2476	25.2730	25.2984	25.3238	25.3492	25.3746
1.000	25.4000	...	..	...	...	...	...	...	...	...

Use previous table to obtain whole inch equivalents to add to decimal equivalents above. All values given in this table are exact; figures to the right of the last place figures are all zeros.

Millimetres to Inches
(Based on 1 inch = 25.4 millimetres, exactly)

Millimetres	0	1	2	3	4	5	6	7	8	9
	Inches									
0	...	0.03937	0.07874	0.11811	0.15748	0.19685	0.23622	0.27559	0.31496	0.35433
10	0.39370	0.43307	0.47244	0.51181	0.55118	0.59055	0.62992	0.66929	0.70866	0.74803
20	0.78740	0.82677	0.86614	0.90551	0.94488	0.98425	1.02362	1.06299	1.10236	1.14173
30	1.18110	1.22047	1.25984	1.29921	1.33858	1.37795	1.41732	1.45669	1.49606	1.53543
40	1.57480	1.61417	1.65354	1.69291	1.73228	1.77165	1.81102	1.85039	1.88976	1.92913
50	1.96850	2.00787	2.04724	2.08661	2.12598	2.16535	2.20472	2.24409	2.28346	2.32283
60	2.36220	2.40157	2.44094	2.48031	2.51969	2.55906	2.59843	2.63780	2.67717	2.71654
70	2.75591	2.79528	2.83465	2.87402	2.91339	2.95276	2.99213	3.03150	3.07087	3.11024
80	3.14961	3.18898	3.22835	3.26772	3.30709	3.34646	3.38583	3.42520	3.46457	3.50394
90	3.54331	3.58268	3.62205	3.66142	3.70079	3.74016	3.77953	3.81890	3.85827	3.89764
100	3.93701	3.97638	4.01575	4.05512	4.09449	4.13386	4.17323	4.21260	4.25197	4.29134
110	4.33071	4.37008	4.40945	4.44882	4.48819	4.52756	4.56693	4.60630	4.64567	4.68504
120	4.72441	4.76378	4.80315	4.84252	4.88189	4.92126	4.96063	5.00000	5.03937	5.07874
130	5.11811	5.15748	5.19685	5.23622	5.27559	5.31496	5.35433	5.39370	5.43307	5.47244
140	5.51181	5.55118	5.59055	5.62992	5.66929	5.70866	5.74803	5.78740	5.82677	5.86614
150	5.90551	5.94488	5.98425	6.02362	6.06299	6.10236	6.14173	6.18110	6.22047	6.25984
160	6.29921	6.33858	6.37795	6.41732	6.45669	6.49606	6.53543	6.57480	6.61417	6.65354
170	6.69291	6.73228	6.77165	6.81102	6.85039	6.88976	6.92913	6.96850	7.00787	7.04724
180	7.08661	7.12598	7.16535	7.20472	7.24409	7.28346	7.32283	7.36220	7.40157	7.44094
190	7.48031	7.51969	7.55906	7.59843	7.63780	7.67717	7.71654	7.75591	7.79528	7.83465
200	7.87402	7.91339	7.95276	7.99213	8.03150	8.07087	8.11024	8.14961	8.18898	8.22835
210	8.26772	8.30709	8.34646	8.38583	8.42520	8.46457	8.50394	8.54331	8.58268	8.62205
220	8.66142	8.70079	8.74016	8.77953	8.81890	8.85827	8.89764	8.93701	8.97638	9.01575
230	9.05512	9.09449	9.13386	9.17323	9.21260	9.25197	9.29134	9.33071	9.37008	9.40945
240	9.44882	9.48819	9.52756	9.56693	9.60630	9.64567	9.68504	9.72441	9.76378	9.80315
250	9.84252	9.88189	9.92126	9.96063	10.0000	10.0394	10.0787	10.1181	10.1575	10.1969
260	10.2362	10.2756	10.3150	10.3543	10.3937	10.4331	10.4724	10.5118	10.5512	10.5906
270	10.6299	10.6693	10.7087	10.7480	10.7874	10.8268	10.8661	10.9055	10.9449	10.9843
280	11.0236	11.0630	11.1024	11.1417	11.1811	11.2205	11.2598	11.2992	11.3386	11.3780
290	11.4173	11.4567	11.4961	11.5354	11.5748	11.6142	11.6535	11.6929	11.7323	11.7717
300	11.8110	11.8504	11.8898	11.9291	11.9685	12.0079	12.0472	12.0866	12.1260	12.1654
310	12.2047	12.2441	12.2835	12.3228	12.3622	12.4016	12.4409	12.4803	12.5197	12.5591
320	12.5984	12.6378	12.6772	12.7165	12.7559	12.7953	12.8346	12.8740	12.9134	12.9528
330	12.9921	13.0315	13.0709	13.1102	13.1496	13.1890	13.2283	13.2677	13.3071	13.3465
340	13.3858	13.4252	13.4646	13.5039	13.5433	13.5827	13.6220	13.6614	13.7008	13.7402
350	13.7795	13.8189	13.8583	13.8976	13.9370	13.9764	14.0157	14.0551	14.0945	14.1339
360	14.1732	14.2126	14.2520	14.2913	14.3307	14.3701	14.4094	14.4488	14.4882	14.5276
370	14.5669	14.6063	14.6457	14.6850	14.7244	14.7638	14.8031	14.8425	14.8819	14.9213
380	14.9606	15.0000	15.0394	15.0787	15.1181	15.1575	15.1969	15.2362	15.2756	15.3150
390	15.3543	15.3937	15.4331	15.4724	15.5118	15.5512	15.5906	15.6299	15.6693	15.7087
400	15.7480	15.7874	15.8268	15.8661	15.9055	15.9449	15.9843	16.0236	16.0630	16.1024
410	16.1417	16.1811	16.2205	16.2598	16.2992	16.3386	16.3780	16.4173	16.4567	16.4961
420	16.5354	16.5748	16.6142	16.6535	16.6929	16.7323	16.7717	16.8110	16.8504	16.8898
430	16.9291	16.9685	17.0079	17.0472	17.0866	17.1260	17.1654	17.2047	17.2441	17.2835
440	17.3228	17.3622	17.4016	17.4409	17.4803	17.5197	17.5591	17.5984	17.6378	17.6772
450	17.7165	17.7559	17.7953	17.8346	17.8740	17.9134	17.9528	17.9921	18.0315	18.0709
460	18.1102	18.1496	18.1890	18.2283	18.2677	18.3071	18.3465	18.3858	18.4252	18.4646
470	18.5039	18.5433	18.5827	18.6220	18.6614	18.7008	18.7402	18.7795	18.8189	18.8583
480	18.8976	18.9370	18.9764	19.0157	19.0551	19.0945	19.1339	19.1732	19.2126	19.2520
490	19.2913	19.3307	19.3701	19.4094	19.4488	19.4882	19.5276	19.5669	19.6063	19.6457

Millimetres to Inches
(Based on 1 inch = 25.4 millimetres, exactly)

Milli-metres	0	1	2	3	4	5	6	7	8	9
	Inches									
500	19.6850	19.7244	19.7638	19.8031	19.8425	19.8819	19.9213	19.9606	20.0000	20.0394
510	20.0787	20.1181	20.1575	20.1969	20.2362	20.2756	20.3150	20.3543	20.3937	20.4331
520	20.4724	20.5118	20.5512	20.5906	20.6299	20.6693	20.7087	20.7480	20.7874	20.8268
530	20.8661	20.9055	20.9449	20.9843	21.0236	21.0630	21.1024	21.1417	21.1811	21.2205
540	21.2598	21.2992	21.3386	21.3780	21.4173	21.4567	21.4961	21.5354	21.5748	21.6142
550	21.6535	21.6929	21.7323	21.7717	21.8110	21.8504	21.8898	21.9291	21.9685	22.0079
560	22.0472	22.0866	22.1260	22.1654	22.2047	22.2441	22.2835	22.3228	22.3622	22.4016
570	22.4409	22.4803	22.5197	22.5591	22.5984	22.6378	22.6772	22.7165	22.7559	22.7953
580	22.8346	22.8740	22.9134	22.9528	22.9921	23.0315	23.0709	23.1102	23.1496	23.1890
590	23.2283	23.2677	23.3071	23.3465	23.3858	23.4252	23.4646	23.5039	23.5433	23.5827
600	23.6220	23.6614	23.7008	23.7402	23.7795	23.8189	23.8583	23.8976	23.9370	23.9764
610	24.0157	24.0551	24.0945	24.1339	24.1732	24.2126	24.2520	24.2913	24.3307	24.3701
620	24.4094	24.4488	24.4882	24.5276	24.5669	24.6063	24.6457	24.6850	24.7244	24.7638
630	24.8031	24.8425	24.8819	24.9213	24.9606	25.0000	25.0394	25.0787	25.1181	25.1575
640	25.1969	25.2362	25.2756	25.3150	25.3543	25.3937	25.4331	25.4724	25.5118	25.5512
650	25.5906	25.6299	25.6693	25.7087	25.7480	25.7874	25.8268	25.8661	25.9055	25.9449
660	25.9843	26.0236	26.0630	26.1024	26.1417	26.1811	26.2205	26.2598	26.2992	26.3386
670	26.3780	26.4173	26.4567	26.4961	26.5354	26.5748	26.6142	26.6535	26.6929	26.7323
680	26.7717	26.8110	26.8504	26.8898	26.9291	26.9685	27.0079	27.0472	27.0866	27.1260
690	27.1654	27.2047	27.2441	27.2835	27.3228	27.3622	27.4016	27.4409	27.4803	27.5197
700	27.5591	27.5984	27.6378	27.6772	27.7165	27.7559	27.7953	27.8346	27.8740	27.9134
710	27.9528	27.9921	28.0315	28.0709	28.1102	28.1496	28.1890	28.2283	28.2677	28.3071
720	28.3465	28.3858	28.4252	28.4646	28.5039	28.5433	28.5827	28.6220	28.6614	28.7008
730	28.7402	28.7795	28.8189	28.8583	28.8976	28.9370	28.9764	29.0157	29.0551	29.0945
740	29.1339	29.1732	29.2126	29.2520	29.2913	29.3307	29.3701	29.4094	29.4488	29.4882
750	29.5276	29.5669	29.6063	29.6457	29.6850	29.7244	29.7638	29.8031	29.8425	29.8819
760	29.9213	29.9606	30.0000	30.0394	30.0787	30.1181	30.1575	30.1969	30.2362	30.2756
770	30.3150	30.3543	30.3937	30.4331	30.4724	30.5118	30.5512	30.5906	30.6299	30.6693
780	30.7087	30.7480	30.7874	30.8268	30.8661	30.9055	30.9449	30.9843	31.0236	31.0630
790	31.1024	31.1417	31.1811	31.2205	31.2598	31.2992	31.3386	31.3780	31.4173	31.4567
800	31.4961	31.5354	31.5748	31.6142	31.6535	31.6929	31.7323	31.7717	31.8110	31.8504
810	31.8898	31.9291	31.9685	32.0079	32.0472	32.0866	32.1260	32.1654	32.2047	32.2441
820	32.2835	32.3228	32.3622	32.4016	32.4409	32.4803	32.5197	32.5591	32.5984	32.6378
830	32.6772	32.7165	32.7559	32.7953	32.8346	32.8740	32.9134	32.9528	32.9921	33.0315
840	33.0709	33.1102	33.1496	33.1890	33.2283	33.2677	33.3071	33.3465	33.3858	33.4252
850	33.4646	33.5039	33.5433	33.5827	33.6220	33.6614	33.7008	33.7402	33.7795	33.8189
860	33.8583	33.8976	33.9370	33.9764	34.0157	34.0551	34.0945	34.1339	34.1732	34.2126
870	34.2520	34.2913	34.3307	34.3701	34.4094	34.4488	34.4882	34.5276	34.5669	34.6063
880	34.6457	34.6850	34.7244	34.7638	34.8031	34.8425	34.8819	34.9213	34.9606	35.0000
890	35.0394	35.0787	35.1181	35.1575	35.1969	35.2362	35.2756	35.3150	35.3543	35.3937
900	35.4331	35.4724	35.5118	35.5512	35.5906	35.6299	35.6693	35.7087	35.7480	35.7874
910	35.8268	35.8661	35.9055	35.9449	35.9843	36.0236	36.0630	36.1024	36.1417	36.1811
920	36.2205	36.2598	36.2992	36.3386	36.3780	36.4173	36.4567	36.4961	36.5354	36.5748
930	36.6142	36.6535	36.6929	36.7323	36.7717	36.8110	36.8504	36.8898	36.9291	36.9685
940	37.0079	37.0472	37.0866	37.1260	37.1654	37.2047	37.2441	37.2835	37.3228	37.3622
950	37.4016	37.4409	37.4803	37.5197	37.5591	37.5984	37.6378	37.6772	37.7165	37.7559
960	37.7953	37.8346	37.8740	37.9134	37.9528	37.9921	38.0315	38.0709	38.1102	38.1496
970	38.1890	38.2283	38.2677	38.3071	38.3465	38.3858	38.4252	38.4646	38.5039	38.5433
980	38.5827	38.6220	38.6614	38.7008	38.7402	38.7795	38.8189	38.8583	38.8976	38.9370
990	38.9764	39.0157	39.0551	39.0945	39.1339	39.1732	39.2126	39.2520	39.2913	39.3307
1000	39.3701	...	...	...	...	...	...	...	...	...

Fractional Inch—Millimetre and Foot—Millimetre Conversion Tables

(Based on 1 inch = 25.4 millimetres, exactly)

FRACTIONAL INCH TO MILLIMETRES

In.	mm.	In.	mm.	In.	mm.	In.	mm.
1/64	0.397	17/64	6.747	33/64	13.097	49/64	19.447
1/32	0.794	9/32	7.144	17/32	13.494	25/32	19.844
3/64	1.191	19/64	7.541	35/64	13.891	51/64	20.241
1/16	1.588	5/16	7.938	9/16	14.288	13/16	20.638
5/64	1.984	21/64	8.334	37/64	14.684	53/64	21.034
3/32	2.381	11/32	8.731	19/32	15.081	27/32	21.431
7/64	2.778	23/64	9.128	39/64	15.478	55/64	21.828
1/8	3.175	3/8	9.525	5/8	15.875	7/8	22.225
9/64	3.572	25/64	9.922	41/64	16.272	57/64	22.622
5/32	3.969	13/32	10.319	21/32	16.669	29/32	23.019
11/64	4.366	27/64	10.716	43/64	17.066	59/64	23.416
3/16	4.762	7/16	11.112	11/16	17.462	15/16	23.812
13/64	5.159	29/64	11.509	45/64	17.859	61/64	24.209
7/32	5.556	15/32	11.906	23/32	18.256	31/32	24.606
15/64	5.953	31/64	12.303	47/64	18.653	63/64	25.003
1/4	6.350	1/2	12.700	3/4	19.050	1	25.400

INCHES TO MILLIMETRES

In.	mm.	In.	mm.	In.	mm.	In.	mm.	In.	mm.	In.	mm.
1	25.4	3	76.2	5	127.0	7	177.8	9	228.6	11	279.4
2	50.8	4	101.6	6	152.4	8	203.2	10	254.0	12	304.8

FEET TO MILLIMETRES

Ft.	mm.	Ft.	mm.	Ft.	mm.	Ft.	mm.	Ft.	mm.
100	30,480	10	3,048	1	304.8	0.1	30.48	0.01	3.048
200	60,960	20	6,096	2	609.6	0.2	60.96	0.02	6.096
300	91,440	30	9,144	3	914.4	0.3	91.44	0.03	9.144
400	121,920	40	12,192	4	1,219.2	0.4	121.92	0.04	12.192
500	152,400	50	15,240	5	1,524.0	0.5	152.40	0.05	15.240
600	182,880	60	18,288	6	1,828.8	0.6	182.88	0.06	18.288
700	213,360	70	21,336	7	2,133.6	0.7	213.36	0.07	21.336
800	243,840	80	24,384	8	2,438.4	0.8	243.84	0.08	24.384
900	274,320	90	27,432	9	2,743.2	0.9	274.32	0.09	27.432
1,000	304,800	100	30,480	10	3,048.0	1.0	304.80	0.10	30.480

Example 1: Find millimetre equivalent of 293 feet, 5 47/64 inches.

200 ft	=	60,960. mm
90 ft	=	27,432. mm
3 ft	=	914.4 mm
5 in.	=	127.0 mm
47/64 in.	=	18.653 mm
293 ft, 5 47/64 in.	=	89,452.053 mm

Example 2: Find millimetre equivalent of 71.86 feet.

70. ft	=	21,336. mm
1. ft	=	304.8 mm
.80 ft	=	243.84 mm
.06 ft	=	18.288 mm
71.86 ft	=	21,902.928 mm

Foot—Metre and Mile—Kilometre Conversion Tables

(Based on 1 foot = 0.3048 metre, exactly)

FEET TO METRES
(1 ft. = 0.3048 m., exactly)

Feet	Metres	Feet	Metres	Feet	Metres	Feet	Metres	Feet	Metres
100	30.480	10	3.048	1	0.305	0.1	0.030	0.01	0.003
200	60.960	20	6.096	2	0.610	0.2	0.061	0.02	0.006
300	91.440	30	9.144	3	0.914	0.3	0.091	0.03	0.009
400	121.920	40	12.192	4	1.219	0.4	0.122	0.04	0.012
500	152.400	50	15.240	5	1.524	0.5	0.152	0.05	0.015
600	182.880	60	18.288	6	1.829	0.6	0.183	0.06	0.018
700	213.360	70	21.336	7	2.134	0.7	0.213	0.07	0.021
800	243.840	80	24.384	8	2.438	0.8	0.244	0.08	0.024
900	274.320	90	27.432	9	2.743	0.9	0.274	0.09	0.027
1,000	304.800	100	30.480	10	3.048	1.0	0.305	0.10	0.030

METRES TO FEET
(1 m. = 3.280840 ft.)

Metres	Feet	Metres	Feet	Metres	Feet	Metres	Feet	Metres	Feet
100	328.084	10	32.808	1	3.281	0.1	0.328	0.01	0.033
200	656.168	20	65.617	2	6.562	0.2	0.656	0.02	0.066
300	984.252	30	98.425	3	9.843	0.3	0.984	0.03	0.098
400	1,312.336	40	131.234	4	13.123	0.4	1.312	0.04	0.131
500	1,640.420	50	164.042	5	16.404	0.5	1.640	0.05	0.164
600	1,968.504	60	196.850	6	19.685	0.6	1.969	0.06	0.197
700	2,296.588	70	229.659	7	22.966	0.7	2.297	0.07	0.230
800	2,624.672	80	262.467	8	26.247	0.8	2.625	0.08	0.262
900	2,952.756	90	295.276	9	29.528	0.9	2.953	0.09	0.295
1,000	3,280.840	100	328.084	10	32.808	1.0	3.281	0.10	0.328

MILES TO KILOMETRES
(1 mile = 1.609344 km., exactly)

Miles	Km.	Miles	Km.	Miles	Km.	Miles	Km.	Miles	Km.
1,000	1,609.34	100	160.93	10	16.09	1	1.61	0.1	0.16
2,000	3,218.69	200	321.87	20	32.19	2	3.22	0.2	0.32
3,000	4,828.03	300	482.80	30	48.28	3	4.83	0.3	0.48
4,000	6,437.38	400	643.74	40	64.37	4	6.44	0.4	0.64
5,000	8,046.72	500	804.67	50	80.47	5	8.05	0.5	0.80
6,000	9,656.06	600	965.61	60	96.56	6	9.66	0.6	0.97
7,000	11,265.41	700	1,126.54	70	112.65	7	11.27	0.7	1.13
8,000	12,874.75	800	1,287.48	80	128.75	8	12.87	0.8	1.29
9,000	14,484.10	900	1,448.41	90	144.84	9	14.48	0.9	1.45
10,000	16,093.44	1,000	1,609.34	100	160.93	10	16.09	1.0	1.61

KILOMETRES TO MILES
(1 km. = 0.6213712 mile)

Km.	Miles	Km.	Miles	Km.	Miles	Km.	Miles	Km.	Miles
1,000	621.37	100	62.14	10	6.21	1	0.62	0.1	0.06
2,000	1,242.74	200	124.27	20	12.43	2	1.24	0.2	0.12
3,000	1,864.11	300	186.41	30	18.64	3	1.86	0.3	0.19
4,000	2,485.48	400	248.55	40	24.85	4	2.49	0.4	0.25
5,000	3,106.86	500	310.69	50	31.07	5	3.11	0.5	0.31
6,000	3,728.23	600	372.82	60	37.28	6	3.73	0.6	0.37
7,000	4,349.60	700	434.96	70	43.50	7	4.35	0.7	0.43
8,000	4,970.97	800	497.10	80	49.71	8	4.97	0.8	0.50
9,000	5,592.34	900	559.23	90	55.92	9	5.59	0.9	0.56
10,000	6,213.71	1,000	621.37	100	62.14	10	6.21	1.0	0.62

Square Inch—Square Centimetre and Square Foot—Square Metre Conversion Tables

(Based on 1 inch = 25.4 centimetres, exactly)

SQUARE INCHES TO SQUARE CENTIMETRES
(1 sq. in. = 6.4516 sq. cm., exactly)

Sq. In.	Sq. Cm.	Sq. In.	Sq. Cm.	Sq. In.	Sq. Cm.	Sq. In.	Sq. Cm.	Sq. In.	Sq. Cm.
100	645.16	10	64.52	1	6.45	0.1	0.65	0.01	0.06
200	1,290.32	20	129.03	2	12.90	0.2	1.29	0.02	0.13
300	1,935.48	30	193.55	3	19.35	0.3	1.94	0.03	0.19
400	2,580.64	40	258.06	4	25.81	0.4	2.58	0.04	0.26
500	3,225.80	50	322.58	5	32.26	0.5	3.23	0.05	0.32
600	3,870.96	60	387.10	6	38.71	0.6	3.87	0.06	0.39
700	4,516.12	70	451.61	7	45.16	0.7	4.52	0.07	0.45
800	5,161.28	80	516.13	8	51.61	0.8	5.16	0.08	0.52
900	5,806.44	90	580.64	9	58.06	0.9	5.81	0.09	0.58
1,000	6,451.60	100	645.16	10	64.52	1.0	6.45	0.10	0.65

SQUARE CENTIMETRES TO SQUARE INCHES
(1 sq. cm. = 0.1550003 sq. in.)

Sq. Cm.	Sq. In.	Sq. Cm.	Sq. In.	Sq. Cm.	Sq. In.	Sq. Cm.	Sq. In.	Sq. Cm.	Sq. In.
100	15.500	10	1.550	1	0.155	0.1	0.016	0.01	0.002
200	31.000	20	3.100	2	0.310	0.2	0.031	0.02	0.003
300	46.500	30	4.650	3	0.465	0.3	0.047	0.03	0.005
400	62.000	40	6.200	4	0.620	0.4	0.062	0.04	0.006
500	77.500	50	7.750	5	0.775	0.5	0.078	0.05	0.008
600	93.000	60	9.300	6	0.930	0.6	0.093	0.06	0.009
700	108.500	70	10.850	7	1.085	0.7	0.109	0.07	0.011
800	124.000	80	12.400	8	1.240	0.8	0.124	0.08	0.012
900	139.500	90	13.950	9	1.395	0.9	0.140	0.09	0.014
1,000	155.000	100	15.500	10	1.550	1.0	0.155	0.10	0.016

SQUARE FEET TO SQUARE METRES
(1 sq. ft. = 0.09290304 sq. m., exactly)

Sq. Ft.	Sq. M.	Sq. Ft.	Sq. M.	Sq. Ft.	Sq. M.	Sq. Ft.	Sq. M.	Sq. Ft.	Sq. M.
1,000	92.903	100	9.290	10	0.929	1	0.093	0.1	0.009
2,000	185.806	200	18.581	20	1.858	2	0.186	0.2	0.019
3,000	278.709	300	27.871	30	2.787	3	0.279	0.3	0.028
4,000	371.612	400	37.161	40	3.716	4	0.372	0.4	0.037
5,000	464.515	500	46.452	50	4.645	5	0.465	0.5	0.046
6,000	557.418	600	55.742	60	5.574	6	0.557	0.6	0.056
7,000	650.321	700	65.032	70	6.503	7	0.650	0.7	0.065
8,000	743.224	800	74.322	80	7.432	8	0.743	0.8	0.074
9,000	836.127	900	83.613	90	8.361	9	0.836	0.9	0.084
10,000	929.030	1,000	92.903	100	9.290	10	0.929	1.0	0.093

SQUARE METRES TO SQUARE FEET
(1 sq. m. = 10.76391 sq. ft.)

Sq. M.	Sq. Ft.	Sq. M.	Sq. Ft.	Sq. M.	Sq. Ft.	Sq. M.	Sq. Ft.	Sq. M.	Sq. Ft.
100	1,076.39	10	107.64	1	10.76	0.1	1.08	0.01	0.11
200	2,152.78	20	215.28	2	21.53	0.2	2.15	0.02	0.22
300	3,229.17	30	322.92	3	32.29	0.3	3.23	0.03	0.32
400	4,305.56	40	430.56	4	43.06	0.4	4.31	0.04	0.43
500	5,381.96	50	538.20	5	53.82	0.5	5.38	0.05	0.54
600	6,458.35	60	645.83	6	64.58	0.6	6.46	0.06	0.65
700	7,534.74	70	753.47	7	75.35	0.7	7.53	0.07	0.75
800	8,611.13	80	861.11	8	86.11	0.8	8.61	0.08	0.86
900	9,687.52	90	968.75	9	96.88	0.9	9.69	0.09	0.97
1,000	10,763.91	100	1,076.39	10	107.64	1.0	10.76	0.10	1.08

Acre — Hectare and U.K. Gallon — Litre Conversion Tables

ACRES TO HECTARES
(1 acre = 0.4046856 hectare)

Acres	0	10	20	30	40	50	60	70	80	90
	Hectares									
0	....	4.047	8.094	12.141	16.187	20.234	24.281	28.328	32.375	36.422
100	40.469	44.515	48.562	52.609	56.656	60.703	64.750	68.797	72.843	76.890
200	80.937	84.984	89.031	93.078	97.125	101.171	105.218	109.265	113.312	117.359
300	121.406	125.453	129.499	133.546	137.593	141.640	145.687	149.734	153.781	157.827
400	161.874	165.921	169.968	174.015	178.062	182.109	186.155	190.202	194.249	198.296
500	202.343	206.390	210.437	214.483	218.530	222.577	226.624	230.671	234.718	238.765
600	242.811	246.858	250.905	254.952	258.999	263.046	267.092	271.139	275.186	279.233
700	283.280	287.327	291.374	295.420	299.467	303.514	307.561	311.608	315.655	319.702
800	323.748	327.795	331.842	335.889	339.936	343.983	348.030	352.076	356.123	360.170
900	364.217	368.264	372.311	376.358	380.404	384.451	388.498	392.545	396.592	400.639
1000	404.686	....	....	....	....	....	....	....	....	....

HECTARES TO ACRES
(1 hectare = 2.471054 acres)

Hectares	0	10	20	30	40	50	60	70	80	90
	Acres									
0	....	24.71	49.42	74.13	98.84	123.55	148.26	172.97	197.68	222.39
100	247.11	271.82	296.53	321.24	345.95	370.66	395.37	420.08	444.79	469.50
200	494.21	518.92	543.63	568.34	593.05	617.76	642.47	667.18	691.90	716.61
300	741.32	766.03	790.74	815.45	840.16	864.87	889.58	914.29	939.00	963.71
400	988.42	1013.13	1037.84	1062.55	1087.26	1111.97	1136.68	1161.40	1186.11	1210.82
500	1235.53	1260.24	1284.95	1309.66	1334.37	1359.08	1383.79	1408.50	1433.21	1457.92
600	1482.63	1507.34	1532.05	1556.76	1581.47	1606.19	1630.90	1655.61	1680.32	1705.03
700	1729.74	1754.45	1779.16	1803.87	1828.58	1853.29	1878.00	1902.71	1927.42	1952.13
800	1976.84	2001.55	2026.26	2050.97	2075.69	2100.40	2125.11	2149.82	2174.53	2199.24
900	2223.95	2248.66	2273.37	2298.08	2322.79	2347.50	2372.21	2396.92	2421.63	2446.34
1000	2471.05	....	....	....	....	....	....	....	....	....

U.K. GALLONS TO LITRES
(1 U.K. gallon = 4.546092 litres)

Imp. Gals.	0	1	2	3	4	5	6	7	8	9
	Litres									
0	....	4.546	9.092	13.638	18.184	22.730	27.277	31.823	36.369	40.915
10	45.461	50.007	54.553	59.099	63.645	68.191	72.737	77.284	81.830	86.376
20	90.922	95.468	100.014	104.560	109.106	113.652	118.198	122.744	127.291	131.837
30	136.383	140.929	145.475	150.021	154.567	159.113	163.659	168.205	172.751	177.298
40	181.844	186.390	190.936	195.482	200.028	204.574	209.120	213.666	218.212	222.759
50	227.305	231.851	236.397	240.943	245.489	250.035	254.581	259.127	263.673	268.219
60	272.766	277.312	281.858	286.404	290.950	295.496	300.042	304.588	309.134	313.680
70	318.226	322.773	327.319	331.865	336.411	340.957	345.503	350.049	354.595	359.141
80	363.687	368.233	372.780	377.326	381.872	386.418	390.964	395.510	400.056	404.602
90	409.148	413.694	418.240	422.787	427.333	431.879	436.425	440.971	445.517	450.063
100	454.609	459.155	463.701	468.247	472.794	477.340	481.886	486.432	490.978	495.524

LITRES TO U.K. GALLONS
(1 litre = 0.2199692 U.K. gallons)

Litres	0	1	2	3	4	5	6	7	8	9
	Imperial Gallons									
0	....	0.220	0.440	0.660	0.880	1.100	1.320	1.540	1.760	1.980
10	2.200	2.420	2.640	2.860	3.080	3.300	3.520	3.739	3.959	4.179
20	4.399	4.619	4.839	5.059	5.279	5.499	5.719	5.939	6.159	6.379
30	6.599	6.819	7.039	7.259	7.479	7.699	7.919	8.139	8.359	8.579
40	8.799	9.019	9.239	9.459	9.679	9.899	10.119	10.339	10.559	10.778
50	10.998	11.218	11.438	11.658	11.878	12.098	12.318	12.538	12.758	12.978
60	13.198	13.418	13.638	13.858	14.078	14.298	14.518	14.738	14.958	15.178
70	15.398	15.618	15.838	16.058	16.278	16.498	16.718	16.938	17.158	17.378
80	17.598	17.818	18.037	18.257	18.477	18.697	18.917	19.137	19.357	19.577
90	19.797	20.017	20.237	20.457	20.677	20.897	21.117	21.337	21.557	21.777
100	21.997	22.217	22.437	22.657	22.877	23.097	23.317	23.537	23.757	23.977

Cubic Inch—Cubic Centimetre and Cubic Foot—Cubic Metre Conversion Tables

(Based on 1 inch = 2.54 centimetres, exactly)

CUBIC INCHES TO CUBIC CENTIMETRES
(1 cu. in. = 16.38706 cu. cm.)

Cu. In.	Cu. Cm.	Cu. In.	Cu. Cm.	Cu. In.	Cu. Cm.	Cu. In.	Cu. Cm.	Cu. In.	Cu. Cm.
100	1,638.71	10	163.87	1	16.39	0.1	1.64	0.01	0.16
200	3,277.41	20	327.74	2	32.77	0.2	3.28	0.02	0.33
300	4,916.12	30	491.61	3	49.16	0.3	4.92	0.03	0.49
400	6,554.82	40	655.48	4	65.55	0.4	6.55	0.04	0.66
500	8,193.53	50	819.35	5	81.94	0.5	8.19	0.05	0.82
600	9,832.24	60	983.22	6	98.32	0.6	9.83	0.06	0.98
700	11,470.94	70	1,147.09	7	114.71	0.7	11.47	0.07	1.15
800	13,109.65	80	1,310.96	8	131.10	0.8	13.11	0.08	1.31
900	14,748.35	90	1,474.84	9	147.48	0.9	14.75	0.09	1.47
1,000	16,387.06	100	1,638.71	10	163.87	1.0	16.39	0.10	1.64

CUBIC CENTIMETRES TO CUBIC INCHES
(1 cu. cm. = 0.06102376 cu. in.)

Cu. Cm.	Cu. In.	Cu. Cm.	Cu. In.	Cu. Cm.	Cu. In.	Cu. Cm.	Cu. In.	Cu. Cm.	Cu. In.
1,000	61.024	100	6.102	10	0.610	1	0.061	0.1	0.006
2,000	122.048	200	12.205	20	1.220	2	0.122	0.2	0.012
3,000	183.071	300	18.307	30	1.831	3	0.183	0.3	0.018
4,000	244.095	400	24.410	40	2.441	4	0.244	0.4	0.024
5,000	305.119	500	30.512	50	3.051	5	0.305	0.5	0.031
6,000	366.143	600	36.614	60	3.661	6	0.366	0.6	0.037
7,000	427.166	700	42.717	70	4.272	7	0.427	0.7	0.043
8,000	488.190	800	48.819	80	4.882	8	0.488	0.8	0.049
9,000	549.214	900	54.921	90	5.492	9	0.549	0.9	0.055
10,000	610.238	1,000	61.024	100	6.102	10	0.610	1.0	0.061

CUBIC FEET TO CUBIC METRES
(1 cu. ft. = 0.02831685 cu. m.)

Cu. Ft.	Cu. M.	Cu. Ft.	Cu. M.	Cu. Ft.	Cu. M.	Cu. Ft.	Cu. M.	Cu. Ft.	Cu. M.
1,000	28.317	100	2.832	10	0.283	1	0.028	0.1	0.003
2,000	56.634	200	5.663	20	0.566	2	0.057	0.2	0.006
3,000	84.951	300	8.495	30	0.850	3	0.085	0.3	0.008
4,000	113.267	400	11.327	40	1.133	4	0.113	0.4	0.011
5,000	141.584	500	14.158	50	1.416	5	0.142	0.5	0.014
6,000	169.901	600	16.990	60	1.699	6	0.170	0.6	0.017
7,000	198.218	700	19.822	70	1.982	7	0.198	0.7	0.020
8,000	226.535	800	22.653	80	2.265	8	0.227	0.8	0.023
9,000	254.852	900	25.485	90	2.549	9	0.255	0.9	0.025
10,000	283.168	1,000	28.317	100	2.832	10	0.283	1.0	0.028

CUBIC METRES TO CUBIC FEET
(1 cu. m. = 35.31466 cu. ft.)

Cu. M.	Cu. Ft.	Cu. M.	Cu. Ft.	Cu. M.	Cu. Ft.	Cu. M.	Cu. Ft.	Cu. M.	Cu. Ft.
100	3,531.47	10	353.15	1	35.31	0.1	3.53	0.01	0.35
200	7,062.93	20	706.29	2	70.63	0.2	7.06	0.02	0.71
300	10,594.40	30	1,059.44	3	105.94	0.3	10.59	0.03	1.06
400	14,125.86	40	1,412.59	4	141.26	0.4	14.13	0.04	1.41
500	17,657.33	50	1,765.73	5	176.57	0.5	17.66	0.05	1.77
600	21,188.80	60	2,118.88	6	211.89	0.6	21.19	0.06	2.12
700	24,720.26	70	2,472.03	7	247.20	0.7	24.72	0.07	2.47
800	28,251.73	80	2,825.17	8	282.52	0.8	28.25	0.08	2.83
900	31,783.19	90	3,178.32	9	317.83	0.9	31.78	0.09	3.18
1,000	35,314.66	100	3,531.47	10	353.15	1.0	35.31	0.10	3.53

Cubic Foot—Litre and Gallon—Litre Conversion Tables

(Based on 1 litre = 1000 cubic centimetres)

CUBIC FEET TO LITRES
(1 cu. ft. = 28.31685 litres)

Cu. Ft.	Litres	Cu. Ft.	Litres	Cu. Ft.	Litres	Cu. Ft.	Litres	Cu. Ft.	Litres
100	2,831.68	10	283.17	1	28.32	0.1	2.83	0.01	0.28
200	5,663.37	20	566.34	2	56.63	0.2	5.66	0.02	0.57
300	8,495.06	30	849.51	3	84.95	0.3	8.50	0.03	0.85
400	11,326.74	40	1,132.67	4	113.27	0.4	11.33	0.04	1.13
500	14,158.42	50	1,415.84	5	141.58	0.5	14.16	0.05	1.42
600	16,990.11	60	1,699.01	6	169.90	0.6	16.99	0.06	1.70
700	19,821.80	70	1,982.18	7	198.22	0.7	19.82	0.07	1.98
800	22,653.48	80	2,263.35	8	226.53	0.8	22.65	0.08	2.27
900	25,485.16	90	2,548.52	9	254.85	0.9	25.49	0.09	2.55
1,000	28,316.85	100	2,831.68	10	283.17	1.0	28.32	0.10	2.83

LITRES TO CUBIC FEET
(1 litre = 0.03531466 cu. ft.)

Litres	Cu. Ft.	Litres	Cu. Ft.	Litres	Cu. Ft.	Litres	Cu. Ft.	Litres	Cu. Ft.
1,000	35.315	100	3.531	10	0.353	1	0.035	0.1	0.004
2,000	70.629	200	7.063	20	0.706	2	0.071	0.2	0.007
3,000	105.944	300	10.594	30	1.059	3	0.106	0.3	0.011
4,000	141.259	400	14.126	40	1.413	4	0.141	0.4	0.014
5,000	176.573	500	17.657	50	1.766	5	0.177	0.5	0.018
6,000	211.888	600	21.189	60	2.119	6	0.212	0.6	0.021
7,000	247.203	700	24.720	70	2.472	7	0.247	0.7	0.025
8,000	282.517	800	28.252	80	2.825	8	0.283	0.8	0.028
9,000	317.832	900	31.783	90	3.178	9	0.318	0.9	0.032
10,000	353.147	1,000	35.315	100	3.531	10	0.353	1.0	0.035

U. S. GALLONS TO LITRES
(1 U. S. gallon = 3.785412 litres)

Gals.	Litres	Gals.	Litres	Gals.	Litres	Gals.	Litres	Gals.	Litres
1,000	3,785.41	100	378.54	10	37.85	1	3.79	0.1	0.38
2,000	7,570.82	200	757.08	20	75.71	2	7.57	0.2	0.76
3,000	11,356.24	300	1,135.62	30	113.56	3	11.36	0.3	1.14
4,000	15,141.65	400	1,514.16	40	151.42	4	15.14	0.4	1.51
5,000	18,927.06	500	1,892.71	50	189.27	5	18.93	0.5	1.89
6,000	22,712.47	600	2,271.25	60	227.12	6	22.71	0.6	2.27
7,000	26,497.88	700	2,649.79	70	264.98	7	26.50	0.7	2.65
8,000	30,283.30	800	3,028.33	80	302.83	8	30.28	0.8	3.03
9,000	34,068.71	900	3,406.87	90	340.69	9	34.07	0.9	3.41
10,000	37,854.12	1,000	3,785.41	100	378.54	10	37.85	1.0	3.79

LITRES TO U. S. GALLONS
(1 litre = 0.2641720 U. S. gallon)

Litres	Gals.	Litres	Gals.	Litres	Gals.	Litres	Gals.	Litres	Gals.
1,000	264.17	100	26.42	10	2.64	1	0.26	0.1	0.03
2,000	528.34	200	52.83	20	5.28	2	0.53	0.2	0.05
3,000	792.52	300	79.25	30	7.93	3	0.79	0.3	0.08
4,000	1,056.69	400	105.67	40	10.57	4	1.06	0.4	0.11
5,000	1,320.86	500	132.09	50	13.21	5	1.32	0.5	0.13
6,000	1,585.03	600	158.50	60	15.85	6	1.59	0.6	0.16
7,000	1,849.20	700	184.92	70	18.49	7	1.85	0.7	0.18
8,000	2,113.38	800	211.34	80	21.13	8	2.11	0.8	0.21
9,000	2,377.55	900	237.75	90	23.78	9	2.38	0.9	0.24
10,000	2,641.72	1,000	264.17	100	26.42	10	2.64	1.0	0.26

Pound—Kilogram and Ounce—Gram Conversion Tables

POUNDS TO KILOGRAMS (1 pound = 0.4535924 kilogram)

Lb.	Kg.	Lb.	Kg.	Lb.	Kg.	Lb.	Kg.	Lb.	Kg.
1,000	453.59	100	45.36	10	4.54	1	0.45	0.1	0.05
2,000	907.18	200	90.72	20	9.07	2	0.91	0.2	0.09
3,000	1,360.78	300	136.08	30	13.61	3	1.36	0.3	0.14
4,000	1,814.37	400	181.44	40	18.14	4	1.81	0.4	0.18
5,000	2,267.96	500	226.80	50	22.68	5	2.27	0.5	0.23
6,000	2,721.55	600	272.16	60	27.22	6	2.72	0.6	0.27
7,000	3,175.15	700	317.51	70	31.75	7	3.18	0.7	0.32
8,000	3,628.74	800	362.87	80	36.29	8	3.63	0.8	0.36
9,000	4,082.33	900	408.23	90	40.82	9	4.08	0.9	0.41
10,000	4,535.92	1,000	453.59	100	45.36	10	4.54	1.0	0.45

KILOGRAMS TO POUNDS (1 kilogram = 2.204622 pounds)

Kg.	Lb.	Kg.	Lb.	Kg.	Lb.	Kg.	Lb.	Kg.	Lb.
1,000	2,204.62	100	220.46	10	22.05	1	2.20	0.1	0.22
2,000	4,409.24	200	440.92	20	44.09	2	4.41	0.2	0.44
3,000	6,613.87	300	661.39	30	66.14	3	6.61	0.3	0.66
4,000	8,818.49	400	881.85	40	88.18	4	8.82	0.4	0.88
5,000	11,023.11	500	1,102.31	50	110.23	5	11.02	0.5	1.10
6,000	13,227.73	600	1,322.77	60	132.28	6	13.23	0.6	1.32
7,000	15,432.35	700	1,543.24	70	154.32	7	15.43	0.7	1.54
8,000	17,636.98	800	1,763.70	80	176.37	8	17.64	0.8	1.76
9,000	19,841.60	900	1,984.16	90	198.42	9	19.84	0.9	1.98
10,000	22,046.22	1,000	2,204.62	100	220.46	10	22.05	1.0	2.20

OUNCES TO GRAMS (1 ounce = 28.34952 grams)

Oz.	G.	Oz.	G.	Oz.	G.	Oz.	G.	Oz.	G.
10	283.50	1	28.35	0.1	2.83	0.01	0.28	0.001	0.03
20	566.99	2	56.70	0.2	5.67	0.02	0.57	0.002	0.06
30	850.49	3	85.05	0.3	8.50	0.03	0.85	0.003	0.09
40	1,133.98	4	113.40	0.4	11.34	0.04	1.13	0.004	0.11
50	1,417.48	5	141.75	0.5	14.17	0.05	1.42	0.005	0.14
60	1,700.97	6	170.10	0.6	17.01	0.06	1.70	0.006	0.17
70	1,984.47	7	198.45	0.7	19.84	0.07	1.98	0.007	0.20
80	2,267.96	8	226.80	0.8	22.68	0.08	2.27	0.008	0.23
90	2,551.46	9	255.15	0.9	25.51	0.09	2.55	0.009	0.26
100	2,834.95	10	283.50	1.0	28.35	0.10	2.83	0.010	0.28

GRAMS TO OUNCES (1 gram = 0.03527397 ounce)

G.	Oz.	G.	Oz.	G.	Oz.	G.	Oz.	G.	Oz.
100	3.527	10	0.353	1	0.035	0.1	0.004	0.01	0.000
200	7.055	20	0.705	2	0.071	0.2	0.007	0.02	0.001
300	10.582	30	1.058	3	0.106	0.3	0.011	0.03	0.001
400	14.110	40	1.411	4	0.141	0.4	0.014	0.04	0.001
500	17.637	50	1.764	5	0.176	0.5	0.018	0.05	0.002
600	21.164	60	2.116	6	0.212	0.6	0.021	0.06	0.002
700	24.692	70	2.469	7	0.247	0.7	0.025	0.07	0.002
800	28.219	80	2.822	8	0.282	0.8	0.028	0.08	0.003
900	31.747	90	3.175	9	0.317	0.9	0.032	0.09	0.003
1,000	35.274	100	3.527	10	0.353	1.0	0.035	0.10	0.004

Pounds per Square Inch—Kilograms per Square Centimetre and Pounds per Square Foot—Kilograms per Square Metre Conversion Tables

POUNDS PER SQUARE INCH TO KILOGRAMS PER SQUARE CENTIMETRE
(1 lb. per sq. in. = 0.07030697 kg. per sq. cm.)

Lb/in.²	Kg/cm²	Lb/in.²	Kg/cm²	Lb/in.²	Kg/cm²	Lb/in.²	Kg/cm²	Lb/in.²	Kg/cm²
1,000	70.307	100	7.031	10	0.703	1	0.070	0.1	0.007
2,000	140.614	200	14.061	20	1.406	2	0.141	0.2	0.014
3,000	210.921	300	21.092	30	2.109	3	0.211	0.3	0.021
4,000	281.228	400	28.123	40	2.812	4	0.281	0.4	0.028
5,000	351.535	500	35.153	50	3.515	5	0.352	0.5	0.035
6,000	421.842	600	42.184	60	4.218	6	0.422	0.6	0.042
7,000	492.149	700	49.215	70	4.921	7	0.492	0.7	0.049
8,000	562.456	800	56.246	80	5.625	8	0.562	0.8	0.056
9,000	632.763	900	63.276	90	6.328	9	0.633	0.9	0.063
10,000	703.070	1,000	70.307	100	7.031	10	0.703	1.0	0.070

KILOGRAMS PER SQUARE CENTIMETRE TO POUNDS PER SQUARE INCH
(1 kg. per sq. cm. = 14.22334 lb. per sq. in.)

Kg/cm²	Lb/in.²	Kg/cm²	Lb/in.²	Kg/cm²	Lb/in.²	Kg/cm²	Lb/in.²	Kg/cm²	Lb/in.²
100	1,422.33	10	142.23	1	14.22	0.1	1.42	0.01	0.14
200	2,844.67	20	284.47	2	28.45	0.2	2.84	0.02	0.28
300	4,267.00	30	426.70	3	42.67	0.3	4.27	0.03	0.43
400	5,689.34	40	568.93	4	56.89	0.4	5.69	0.04	0.57
500	7,111.67	50	711.17	5	71.12	0.5	7.11	0.05	0.71
600	8,534.00	60	853.40	6	85.34	0.6	8.53	0.06	0.85
700	9,956.34	70	995.63	7	99.56	0.7	9.96	0.07	1.00
800	11,378.67	80	1,137.87	8	113.79	0.8	11.38	0.08	1.14
900	12,801.01	90	1,280.10	9	128.01	0.9	12.80	0.09	1.28
1,000	14,223.34	100	1,422.33	10	142.23	1.0	14.22	0.10	1.42

POUNDS PER SQUARE FOOT TO KILOGRAMS PER SQUAPE METRE
(1 lb. per sq. ft. = 4.882429 kg. per sq. m.)

Lb/ft²	Kg/m²	Lb/ft²	Kg/m²	Lb/ft²	Kg/m²	Lb/ft²	Kg/m²	Lb/ft²	Kg/m²
1,000	4,882.43	100	488.24	10	48.82	1	4.88	0.1	0.49
2,000	9,764.86	200	976.49	20	97.65	2	9.76	0.2	0.98
3,000	14,647.29	300	1,464.73	30	146.47	3	14.65	0.3	1.46
4,000	19,529.72	400	1,952.97	40	195.30	4	19.53	0.4	1.95
5,000	24,412.14	500	2,441.21	50	244.12	5	24.41	0.5	2.44
6,000	29,294.57	600	2,929.46	60	292.95	6	29.29	0.6	2.93
7,000	34,177.00	700	3,417.70	70	341.77	7	34.18	0.7	3.42
8,000	39,059.43	800	3,905.94	80	390.59	8	39.06	0.8	3.91
9,000	43,941.86	900	4,394.19	90	439.42	9	43.94	0.9	4.39
10,000	48,824.28	1,000	4,882.43	100	488.24	10	48.82	1.0	4.88

KILOGRAMS PER SQUARE METRE TO POUNDS PER SQUARE FOOT
(1 kg. per sq. m. = 0.2048161 lb. per sq. ft.)

Kg/m²	Lb/ft²	Kg/m²	Lb/ft²	Kg/m²	Lb/ft²	Kg/m²	Lb/ft²	Kg/m²	Lb/ft²
1,000	204.82	100	20.48	10	2.05	1	0.20	0.1	0.02
2,000	409.63	200	40.96	20	4.10	2	0.41	0.2	0.04
3,000	614.45	300	61.44	30	6.14	3	0.61	0.3	0.06
4,000	819.26	400	81.93	40	8.19	4	0.82	0.4	0.08
5,000	1,024.08	500	102.41	50	10.24	5	1.02	0.5	0.10
6,000	1,228.90	600	122.89	60	12.29	6	1.23	0.6	0.12
7,000	1,433.71	700	143.37	70	14.34	7	1.43	0.7	0.14
8,000	1,638.53	800	163.85	80	16.39	8	1.64	0.8	0.16
9,000	1,843.34	900	184.33	90	18.43	9	1.84	0.9	0.18
10,000	2,048.16	1,000	204.82	100	20.48	10	2.05	1.0	0.20

Pounds per Square Inch — Kilonewtons per Square Metre and Tons per Square Inch — Hectobars Conversion Tables*

POUNDS PER SQ. IN. TO KILONEWTONS PER SQ. METRE
(1 lb. per sq. in. = 6.894757 kilonewtons per sq. metre)

Lb. per Sq. In.	0	1	2	3	4	5	6	7	8	9
	Kilonewtons per Sq. Metre									
0	....	6.895	13.790	20.684	27.579	34.474	41.369	48.263	55.158	62.053
10	68.948	75.842	82.737	89.632	96.527	103.421	110.316	117.211	124.106	131.000
20	137.895	144.790	151.685	158.579	165.474	172.369	179.264	186.158	193.053	199.948
30	206.843	213.737	220.632	227.527	234.422	241.316	248.211	255.106	262.001	268.896
40	275.790	282.685	289.580	296.475	303.369	310.264	317.159	324.054	330.948	337.843
50	344.738	351.633	358.527	365.422	372.317	379.212	386.106	393.001	399.896	406.791
60	413.685	420.580	427.475	434.370	441.264	448.159	455.054	461.949	468.843	475.738
70	482.633	489.528	496.423	503.317	510.212	517.107	524.002	530.896	537.791	544.686
80	551.581	558.475	565.370	572.265	579.160	586.054	592.949	599.844	606.739	613.633
90	620.528	627.423	634.318	641.212	648.107	655.002	661.897	668.791	675.686	682.581
100	689.476	696.370	703.265	710.160	717.055	723.949	730.844	737.739	744.634	751.529

KILONEWTONS PER SQ. METRE TO POUNDS PER SQ. IN.
(1 kilonewton per sq. metre = 0.1450377 lb. per sq. in.)

kN per Sq. M.	0	1	2	3	4	5	6	7	8	9
	Pounds per Sq. Inch									
0	....	0.145	0.290	0.435	0.580	0.725	0.870	1.015	1.160	1.305
10	1.450	1.595	1.740	1.885	2.031	2.176	2.321	2.466	2.611	2.756
20	2.901	3.046	3.191	3.336	3.481	3.626	3.771	3.916	4.061	4.206
30	4.351	4.496	4.641	4.786	4.931	5.076	5.221	5.366	5.511	5.656
40	5.802	5.947	6.092	6.237	6.382	6.527	6.672	6.817	6.962	7.107
50	7.252	7.397	7.542	7.687	7.832	7.977	8.122	8.267	8.412	8.557
60	8.702	8.847	8.992	9.137	9.282	9.427	9.572	9.718	9.863	10.008
70	10.153	10.298	10.443	10.588	10.733	10.878	11.023	11.168	11.313	11.458
80	11.603	11.748	11.893	12.038	12.183	12.328	12.473	12.618	12.763	12.908
90	13.053	13.198	13.343	13.489	13.634	13.779	13.924	14.069	14.214	14.359
100	14.504	14.649	14.794	14.939	15.084	15.229	15.374	15.519	15.664	15.809

TONS PER SQ. IN. TO HECTOBARS
(1 ton per sq. in. = 1.544426 hectobars)

Tons per Sq. In.	0	1	2	3	4	5	6	7	8	9
	Hectobars									
0	....	1.54	3.09	4.63	6.18	7.72	9.27	10.81	12.36	13.90
10	15.44	16.99	18.53	20.08	21.62	23.17	24.71	26.26	27.80	29.34
20	30.89	32.43	33.98	35.52	37.07	38.61	40.16	41.70	43.24	44.79
30	46.33	47.88	49.42	50.97	52.51	54.06	55.60	57.14	58.69	60.23
40	61.78	63.32	64.87	66.41	67.95	69.50	71.04	72.59	74.13	75.68
50	77.22	78.77	80.31	81.85	83.40	84.94	86.49	88.03	89.58	91.12
60	92.67	94.21	95.75	97.30	98.84	100.39	101.93	103.48	105.02	106.57
70	108.11	109.65	111.20	112.74	114.29	115.83	117.38	118.92	120.47	122.01
80	123.55	125.10	126.64	128.19	129.73	131.28	132.82	134.37	135.91	137.45
90	139.00	140.54	142.09	143.63	145.18	146.72	148.26	149.81	151.35	152.90
100	154.44	155.99	157.53	159.08	160.62	162.16	163.71	165.25	166.80	168.34

HECTOBARS TO TONS PER SQ. IN.
(1 hectobar = 0.6474898 tons per sq. in.)

Hecto-bars	0	1	2	3	4	5	6	7	8	9
	Tons per Square Inch									
0	....	0.65	1.29	1.94	2.59	3.24	3.88	4.53	5.18	5.83
10	6.47	7.12	7.77	8.42	9.06	9.71	10.36	11.01	11.65	12.30
20	12.95	13.60	14.24	14.89	15.54	16.19	16.83	17.48	18.13	18.78
30	19.42	20.07	20.72	21.37	22.01	22.66	23.31	23.96	24.60	25.25
40	25.90	26.55	27.19	27.84	28.49	29.14	29.78	30.43	31.08	31.73
50	32.37	33.02	33.67	34.32	34.96	35.61	36.26	36.91	37.55	38.20
60	38.85	39.50	40.14	40.79	41.44	42.09	42.73	43.38	44.03	44.68
70	45.32	45.97	46.62	47.27	47.91	48.56	49.21	49.86	50.50	51.15
80	51.80	52.45	53.09	53.74	54.39	55.04	55.68	56.33	56.98	57.63
90	58.27	58.92	59.57	60.22	60.86	61.51	62.16	62.81	63.45	64.10
100	64.75	65.40	66.04	66.69	67.34	67.99	68.63	69.28	69.93	70.58

* *Note:* 1 kilonewton/metre^2 = 1 kilopascal; 1 hectobar = 10,000 kilopascals.

Pounds per Cubic Inch—Grams per Cubic Centimetre and Pounds per Cubic Foot—Kilograms per Cubic Metre Conversion Tables

POUNDS PER CUBIC INCH TO GRAMS PER CUBIC CENTIMETRE
(1 lb. per cu. in. = 27.67990 g. per cu. cm.)

Lb./in.³	G./cm.³	Lb./in.³	G./cm.³	Lb./in.³	G./cm.³	Lb./in.³	G./cm.³	Lb./in.³	G./cm.³
100	2,767.99	10	276.80	1	27.68	0.1	2.77	0.01	0.28
200	5,535.98	20	553.60	2	55.36	0.2	5.54	0.02	0.55
300	8,303.97	30	830.40	3	83.04	0.3	8.30	0.03	0.83
400	11,071.96	40	1,107.20	4	110.72	0.4	11.07	0.04	1.11
500	13,839.95	50	1,384.00	5	138.40	0.5	13.84	0.05	1.38
600	16,607.94	60	1,660.79	6	166.08	0.6	16.61	0.06	1.66
700	19,375.93	70	1,937.59	7	193.76	0.7	19.38	0.07	1.94
800	22,143.92	80	2,214.39	8	221.44	0.8	22.14	0.08	2.21
900	24,911.91	90	2,491.19	9	249.12	0.9	24.91	0.09	2.49
1,000	27,679.90	100	2,767.99	10	276.80	1.0	27.68	0.10	2.77

GRAMS PER CUBIC CENTIMETRE TO POUNDS PER CUBIC INCH
(1 g. per cu. cm. = 0.03612730 lb. per cu. in.)

G./cm.³	Lb./in.³	G./cm.³	Lb./in.³	G./cm.³	Lb./in.³	G./cm.³	Lb./in.³	G./cm.³	Lb./in.³
1,000	36.127	100	3.613	10	0.361	1	0.036	0.1	0.004
2,000	72.255	200	7.225	20	0.723	2	0.072	0.2	0.007
3,000	108.382	300	10.838	30	1.084	3	0.108	0.3	0.011
4,000	144.509	400	14.451	40	1.445	4	0.145	0.4	0.014
5,000	180.636	500	18.064	50	1.806	5	0.181	0.5	0.018
6,000	216.764	600	21.676	60	2.168	6	0.217	0.6	0.022
7,000	252.891	700	25.289	70	2.529	7	0.253	0.7	0.025
8,000	289.018	800	28.902	80	2.890	8	0.289	0.8	0.029
9,000	325.146	900	32.515	90	3.251	9	0.325	0.9	0.033
10,000	361.273	1,000	36.127	100	3.613	10	0.361	1.0	0.036

POUNDS PER CUBIC FOOT TO KILOGRAMS PER CUBIC METRE
(1 lb. per cu. ft. = 16.01846 kg. per cu. m.)

Lb./ft.³	Kg./m.³	Lb./ft.³	Kg./m.³	Lb./ft.³	Kg./m.³	Lb./ft.³	Kg./m.³	Lb./ft.³	Kg./m.³
100	1,601.85	10	160.18	1	16.02	0.1	1.60	0.01	0.16
200	3,203.69	20	320.37	2	32.04	0.2	3.20	0.02	0.32
300	4,805.54	30	480.55	3	48.06	0.3	4.81	0.03	0.48
400	6,407.38	40	640.74	4	64.07	0.4	6.41	0.04	0.64
500	8,009.23	50	800.92	5	80.09	0.5	8.01	0.05	0.80
600	9,611.08	60	961.11	6	96.11	0.6	9.61	0.06	0.96
700	11,212.92	70	1,121.29	7	112.13	0.7	11.21	0.07	1.12
800	12,814.77	80	1,281.48	8	128.15	0.8	12.81	0.08	1.28
900	14,416.61	90	1,441.66	9	144.17	0.9	14.42	0.09	1.44
1,000	16,018.46	100	1,601.85	10	160.18	1.0	16.02	0.10	1.60

KILOGRAMS PER CUBIC METRE TO POUNDS PER CUBIC FOOT
(1 kg. per cu. m. = 0.06242797 lb. per cu. ft.)

Kg./m.³	Lb./ft.³	Kg./m.³	Lb./ft.³	Kg./m.³	Lb./ft.³	Kg./m.³	Lb./ft.³	Kg./m.³	Lb./ft.³
1,000	62.428	100	6.243	10	0.624	1	0.062	0.1	0.006
2,000	124.856	200	12.486	20	1.249	2	0.125	0.2	0.012
3,000	187.284	300	18.728	30	1.873	3	0.187	0.3	0.019
4,000	249.712	400	24.971	40	2.497	4	0.250	0.4	0.025
5,000	312.140	500	31.214	50	3.121	5	0.312	0.5	0.031
6,000	374.568	600	37.457	60	3.746	6	0.375	0.6	0.037
7,000	436.996	700	43.700	70	4.370	7	0.437	0.7	0.044
8,000	499.424	800	49.942	80	4.994	8	0.499	0.8	0.050
9,000	561.852	900	56.185	90	5.619	9	0.562	0.9	0.056
10,000	624.280	1,000	62.428	100	6.243	10	0.624	1.0	0.062

British Thermal Unit — Kilojoule and Foot Pound — Joule Conversion Tables

BRITISH THERMAL UNITS TO KILOJOULES

(1 Btu. = 1055.056 joules.)

Btu.	0	100	200	300	400	500	600	700	800	900
	Kilojoules									
0	....	105.51	211.01	316.52	422.02	527.53	633.03	738.54	844.04	949.55
1000	1055.06	1160.56	1266.07	1371.57	1477.08	1582.58	1688.09	1793.60	1899.10	2004.61
2000	2110.11	2215.62	2321.12	2426.63	2532.13	2637.64	2743.15	2848.65	2954.16	3059.66
3000	3165.17	3270.67	3376.18	3481.68	3587.19	3692.70	3798.20	3903.71	4009.21	4114.72
4000	4220.22	4325.73	4431.24	4536.74	4642.25	4747.75	4853.26	4958.76	5064.27	5169.77
5000	5275.28	5380.79	5486.29	5591.80	5697.30	5802.81	5908.31	6013.82	6119.32	6224.83
6000	6330.34	6435.84	6541.35	6646.85	6752.36	6857.86	6963.37	7068.88	7174.38	7279.89
7000	7385.39	7490.90	7596.40	7701.91	7807.41	7912.92	8018.43	8123.93	8229.44	8334.94
8000	8440.45	8545.95	8651.46	8756.96	8862.47	8967.98	9073.48	9178.99	9284.49	9390.00
9000	9495.50	9601.01	9706.52	9812.02	9917.53	10023.0	10128.5	10234.0	10339.5	10445.1
10000	10550.6	....	....	....	....	....	....	....	....	....

KILOJOULES TO BRITISH THERMAL UNITS

(1 joule = 0.0009478170 Btu.)

kJ	0	100	200	300	400	500	600	700	800	900
	British Thermal Units									
0	....	94.78	189.56	284.35	379.13	473.91	568.69	663.47	758.25	853.04
1000	947.82	1042.60	1137.38	1232.16	1326.94	1421.73	1516.51	1611.29	1706.07	1800.85
2000	1895.63	1990.42	2085.20	2179.98	2274.76	2369.54	2464.32	2559.11	2653.89	2748.67
3000	2843.45	2938.23	3033.01	3127.80	3222.58	3317.36	3412.14	3506.92	3601.70	3696.49
4000	3791.27	3886.05	3980.83	4075.61	4170.39	4265.18	4359.96	4454.74	4549.52	4644.30
5000	4739.08	4833.87	4928.65	5023.43	5118.21	5212.99	5307.78	5402.56	5497.34	5592.12
6000	5686.90	5781.68	5876.47	5971.25	6066.03	6160.81	6255.59	6350.37	6445.16	6539.94
7000	6634.72	6729.50	6824.28	6919.06	7013.85	7108.63	7203.41	7298.19	7392.97	7487.75
8000	7582.54	7677.32	7772.10	7866.88	7961.66	8056.44	8151.23	8246.01	8340.79	8435.57
9000	8530.35	8625.13	8719.92	8814.70	8909.48	9004.26	9099.04	9193.82	9288.61	9383.39
10000	9478.17	....	....	....	....	....	....	....	....	....

FOOT POUNDS TO JOULES

(1 foot pound = 1.355818 joules)

Ft. Lbf.	0	1	2	3	4	5	6	7	8	9
	Joules									
0	....	1.356	2.712	4.067	5.423	6.779	8.135	9.491	10.847	12.202
10	13.558	14.914	16.270	17.626	18.981	20.337	21.693	23.049	24.405	25.761
20	27.116	28.472	29.828	31.184	32.540	33.895	35.251	36.607	37.963	39.319
30	40.675	42.030	43.386	44.742	46.098	47.454	48.809	50.165	51.521	52.877
40	54.233	55.589	56.944	58.300	59.656	61.012	62.368	63.723	65.079	66.435
50	67.791	69.147	70.503	71.858	73.214	74.570	75.926	77.282	78.637	79.993
60	81.349	82.705	84.061	85.417	86.772	88.128	89.484	90.840	92.196	93.551
70	94.907	96.263	97.619	98.975	100.331	101.686	103.042	104.398	105.754	107.110
80	108.465	109.821	111.177	112.533	113.889	115.245	116.600	117.956	119.312	120.668
90	122.024	123.379	124.735	126.091	127.447	128.803	130.159	131.514	132.870	134.226
100	135.582	136.938	138.293	139.649	141.005	142.361	143.717	145.073	146.428	147.784

JOULES TO FOOT POUNDS

(1 joule = 0.7375621 foot pound)

J	0	1	2	3	4	5	6	7	8	9
	Foot Pounds									
0	....	0.7376	1.4751	2.2127	2.9502	3.6878	4.4254	5.1629	5.9005	6.6381
10	7.3756	8.1132	8.8507	9.5883	10.3259	11.0634	11.8010	12.5386	13.2761	14.0137
20	14.7512	15.4888	16.2264	16.9639	17.7015	18.4391	19.1766	19.9142	20.6517	21.3893
30	22.1269	22.8644	23.6020	24.3395	25.0771	25.8147	26.5522	27.2898	28.0274	28.7649
40	29.5025	30.2400	30.9776	31.7152	32.4527	33.1903	33.9279	34.6654	35.4030	36.1405
50	36.8781	37.6157	38.3532	39.0908	39.8284	40.5659	41.3035	42.0410	42.7786	43.5162
60	44.2537	44.9913	45.7289	46.4664	47.2040	47.9415	48.6791	49.4167	50.1542	50.8918
70	51.6293	52.3669	53.1045	53.8420	54.5796	55.3172	56.0547	56.7923	57.5298	58.2674
80	59.0050	59.7425	60.4801	61.2177	61.9552	62.6928	63.4303	64.1679	64.9055	65.6430
90	66.3806	67.1182	67.8557	68.5933	69.3308	70.0684	70.8060	71.5435	72.2811	73.0186
100	73.7562	74.4938	75.2313	75.9689	76.7065	77.4440	78.1816	78.9191	79.6567	80.3943

Pound-force — Newton and Horsepower — Kilowatt Conversion Tables

POUNDS-FORCE TO NEWTONS
(1 pound-force = 4.448222 newtons.)

Lbf.	0	1	2	3	4	5	6	7	8	9
	Newtons									
0	....	4.448	8.896	13.345	17.793	22.241	26.689	31.138	35.586	40.034
10	44.482	48.930	53.379	57.827	62.275	66.723	71.172	75.620	80.068	84.516
20	88.964	93.413	97.861	102.309	106.757	111.206	115.654	120.102	124.550	128.998
30	133.447	137.895	142.343	146.791	151.240	155.688	160.136	164.584	169.032	173.481
40	177.929	182.377	186.825	191.274	195.722	200.170	204.618	209.066	213.515	217.963
50	222.411	226.859	231.308	235.756	240.204	244.652	249.100	253.549	257.997	262.445
60	266.893	271.342	275.790	280.238	284.686	289.134	293.583	298.031	302.479	306.927
70	311.376	315.824	320.272	324.720	329.168	333.617	338.065	342.513	346.961	351.410
80	355.858	360.306	364.754	369.202	373.651	378.099	382.547	386.995	391.444	395.892
90	400.340	404.788	409.236	413.685	418.133	422.581	427.029	431.478	435.926	440.374
100	444.822	449.270	453.719	458.167	462.615	467.063	471.512	475.960	480.408	484.856

NEWTONS TO POUND-FORCE
(1 newton = 0.2248089 pound-force.)

N	0	1	2	3	4	5	6	7	8	9
	Pounds-force									
0	....	0.22481	0.44962	0.67443	0.89924	1.12404	1.34885	1.57366	1.79847	2.02328
10	2.24809	2.47290	2.69771	2.92252	3.14732	3.37213	3.59694	3.82175	4.04656	4.27137
20	4.49618	4.72099	4.94580	5.17060	5.39541	5.62022	5.84503	6.06984	6.29465	6.51946
30	6.74427	6.96908	7.19388	7.41869	7.64350	7.86831	8.09312	8.31793	8.54274	8.76755
40	8.99236	9.21716	9.44197	9.66678	9.89159	10.1164	10.3412	10.5660	10.7908	11.0156
50	11.2404	11.4653	11.6901	11.9149	12.1397	12.3645	12.5893	12.8141	13.0389	13.2637
60	13.4885	13.7133	13.9382	14.1630	14.3878	14.6126	14.8374	15.0622	15.2870	15.5118
70	15.7366	15.9614	16.1862	16.4110	16.6359	16.8607	17.0855	17.3103	17.5351	17.7599
80	17.9847	18.2095	18.4343	18.6591	18.8839	19.1088	19.3336	19.5584	19.7832	20.0080
90	20.2328	20.4576	20.6824	20.9072	21.1320	21.3568	21.5817	21.8065	22.0313	22.2561
100	22.4809	22.7057	22.9305	23.1553	23.3801	23.6049	23.8297	24.0546	24.2794	24.5042

HORSEPOWER TO KILOWATTS
(1 hp. = 0.7456999 kw.)†

Hp.	Kw.	Hp.	Kw.	Hp.	Kw.	Hp.	Kw.	Hp.	Kw.
1,000	745.7	100	74.6	10	7.5	1	0.7	0.1	0.1
2,000	1,491.4	200	149.1	20	14.9	2	1.5	0.2	0.1
3,000	2,237.1	300	223.7	30	22.4	3	2.2	0.3	0.2
4,000	2,982.8	400	298.3	40	29.8	4	3.0	0.4	0.3
5,000	3,728.5	500	372.8	50	37.3	5	3.7	0.5	0.4
6,000	4,474.2	600	447.4	60	44.7	6	4.5	0.6	0.4
7,000	5,219.9	700	522.0	70	52.2	7	5.2	0.7	0.5
8,000	5,965.6	800	596.6	80	59.7	8	6.0	0.8	0.6
9,000	6,711.3	900	671.1	90	67.1	9	6.7	0.9	0.7
10,000	7,457.0	1,000	745.7	100	74.6	10	7.5	1.0	0.7

KILOWATTS TO HORSEPOWER
(1 kw. = 1.341022 hp.)†

Kw.	Hp.	Kw.	Hp.	Kw.	Hp.	Kw.	Hp.	Kw.	Hp.
1,000	1,341.0	100	134.1	10	13.4	1	1.3	0.1	0.1
2,000	2,682.0	200	268.2	20	26.8	2	2.7	0.2	0.3
3,000	4,023.1	300	402.3	30	40.2	3	4.0	0.3	0.4
4,000	5,364.1	400	536.4	40	53.6	4	5.4	0.4	0.5
5,000	6,705.1	500	670.5	50	67.1	5	6.7	0.5	0.7
6,000	8,046.1	600	804.6	60	80.5	6	8.0	0.6	0.8
7,000	9,387.2	700	938.7	70	93.9	7	9.4	0.7	0.9
8,000	10,728.2	800	1,072.8	80	107.3	8	10.7	0.8	1.1
9,000	12,069.2	900	1,206.9	90	120.7	9	12.1	0.9	1.2
10,000	13,410.2	1,000	1,341.0	100	134.1	10	13.4	1.0	1.3

† Based on 1 horsepower = 550 foot-pounds per second.

Fahrenheit — Celsius (Centigrade) Conversion. — A simple way to convert a Fahrenheit temperature reading into a Celsius temperature reading or vice versa is to enter the accompanying table in the center or boldface column of figures. These figures refer to the temperature in either Fahrenheit or Celsius degrees. If it is desired to convert from Fahrenheit to Celsius degrees, consider the center column as a table of Fahrenheit temperatures and read the corresponding Celsius temperature in the column at the left. If it is desired to convert from Celsius to Fahrenheit degrees, consider the center column as a table of Celsius values, and read the corresponding Fahrenheit temperature on the right.

Deg.C.		Deg.F.	Deg.C.		Deg.F.
0.56	**1**	1.8	3.33	**6**	10.8
1.11	**2**	3.6	3.89	**7**	12.6
1.67	**3**	5.4	4.44	**8**	14.4
2.22	**4**	7.2	5.00	**9**	16.2
2.78	**5**	9.0	5.56	**10**	18.0

Interpolation factors are given for use with that portion of the table in which the center column advances in increments of 10. To illustrate, suppose it is desired to find the Fahrenheit equivalent of 314 degrees C. The equivalent of 310 degrees C., found in the body of the main table, is seen to be 590.0 degrees F. The Fahrenheit equivalent of a 4-degree C. difference is seen to be 7.2, as read in the table of interpolating factors. The answer is the sum or 597.2 degrees F.

Fahrenheit — Celsius (Centigrade) Conversion Table

Deg.C.		Deg.F.	Deg.C.		Deg.F.	Deg.C.		Deg.F.	Deg.C.		Deg.F.
—273	**—459.4**	...	—101	**—150**	—238	— 8.3	**17**	62.6	9.4	**49**	120.2
—268	**—450**	...	— 96	**—140**	—220	— 7.8	**18**	64.4	10.0	**50**	122.0
—262	**—440**	...	— 90	**—130**	—202	— 7.2	**19**	66.2	10.6	**51**	123.8
—257	**—430**	...	— 84	**—120**	—184	— 6.7	**20**	68.0	11.1	**52**	125.6
—251	**—420**	...	— 79	**—110**	—166	— 6.1	**21**	69.8	11.7	**53**	127.4
—246	**—410**	...	— 73	**—100**	—148	— 5.6	**22**	71.6	12.2	**54**	129.2
—240	**—400**	...	— 68	**— 90**	—130	— 5.0	**23**	73.4	12.8	**55**	131.0
—234	**—390**	...	— 62	**— 80**	—112	— 4.4	**24**	75.2	13.3	**56**	132.8
—229	**—380**	...	— 57	**— 70**	— 94	— 3.9	**25**	77.0	13.9	**57**	134.6
—223	**—370**	...	— 51	**— 60**	— 76	— 3.3	**26**	78.8	14.4	**58**	136.4
—218	**—360**	...	—46	**—50**	—58	— 2.8	**27**	80.6	15.0	**59**	138.2
—212	**—350**	...	—40	**—40**	—40	— 2.2	**28**	82.4	15.6	**60**	140.0
—207	**—340**	...	—34	**—30**	—22	— 1.7	**29**	84.2	16.1	**61**	141.8
—201	**—330**	...	—29	**—20**	— 4	— 1.1	**30**	86.0	16.7	**62**	143.6
—196	**—320**	...	—23	**—10**	14	— 0.6	**31**	87.8	17.2	**63**	145.4
—190	**—310**	...	—17.8	**0**	32—	0—	**32**	89.6	17.8	**64**	147.2
—184	**—300**	...	—17.2	**1**	33.8	0.6	**33**	91.4	18.3	**65**	149.0
—179	**—290**	...	—16.7	**2**	35.6	1.1	**34**	93.2	18.9	**66**	150.8
—173	**—280**	...	—16.1	**3**	37.4	1.7	**35**	95.0	19.4	**67**	152.6
—169	**—273**	—459.4	—15.6	**4**	39.2	2.2	**36**	96.8	20.0	**68**	154.4
—168	**—270**	—454	—15.0	**5**	41.0	2.7	**37**	98.6	20.6	**69**	156.2
—162	**—260**	—436	—14.4	**6**	42.8	3.3	**38**	100.4	21.1	**70**	158.0
—157	**—250**	—418	—13.9	**7**	44.6	3.9	**39**	102.2	21.7	**71**	159.8
—151	**—240**	—400	—13.3	**8**	46.4	4.4	**40**	104.0	22.2	**72**	161.6
—146	**—230**	—382	—12.8	**9**	48.2	5.0	**41**	105.8	22.8	**73**	163.4
—140	**—220**	—364	—12.2	**10**	50.0	5.6	**42**	107.6	23.3	**74**	165.2
—134	**—210**	—346	—11.7	**11**	51.8	6.1	**43**	109.4	23.9	**75**	167.0
—129	**—200**	—328	—11.1	**12**	53.6	6.7	**44**	111.2	24.4	**76**	168.8
—123	**—190**	—310	—10.6	**13**	55.4	7.2	**45**	113.0	25.0	**77**	170.6
—118	**—180**	—292	—10.0	**14**	57.2	7.8	**46**	114.8	25.6	**78**	172.4
—112	**—170**	—274	— 9.4	**15**	59.0	8.3	**47**	116.6	26.1	**79**	174.2
—107	**—160**	—256	— 8.9	**16**	60.8	8.9	**48**	118.4	26.7	**80**	176.0

Fahrenheit — Celsius (Centigrade) Conversion Table (*Continued*)

Deg.C.		Deg.F.	Deg.C.		Deg.F.	Deg.C.		Deg.F.	Deg.C.		Deg.F.
27.2	81	177.8	58.3	137	278.6	89.4	193	379.4	304.4	580	1076
27.8	82	179.6	58.9	138	280.4	90.0	194	381.2	310.0	590	1094
28.3	83	181.4	59.4	139	282.2	90.6	195	383.0	315.6	600	1112
28.9	84	183.2	60.0	140	284.0	91.1	196	384.8	321.1	610	1130
29.4	85	185.0	60.6	141	285.8	91.7	197	386.6	326.7	620	1148
30.0	86	186.8	61.1	142	287.6	92.2	198	388.4	332.2	630	1166
30.6	87	188.6	61.7	143	289.4	92.8	199	390.2	337.8	640	1184
31.1	88	190.4	62.2	144	291.2	93.3	200	392.0	343.3	650	1202
31.7	89	192.2	62.8	145	293.0	93.9	201	393.8	348.9	660	1220
32.2	90	194.0	63.3	146	294.8	94.4	202	395.6	354.4	670	1238
32.8	91	195.8	63.9	147	296.6	95.0	203	397.4	360.0	680	1256
33.3	92	197.6	64.4	148	298.4	95.6	204	399.2	365.6	690	1274
33.9	93	199.4	65.0	149	300.2	96.1	205	401.0	371.1	700	1292
34.4	94	201.2	65.6	150	302.0	96.7	206	402.8	376.7	710	1310
35.0	95	203.0	66.1	151	303.8	97.2	207	404.6	382.2	720	1328
35.6	96	204.8	66.7	152	305.6	97.8	208	406.4	387.8	730	1346
36.1	97	206.6	67.2	153	307.4	98.3	209	408.2	393.3	740	1364
36.7	98	208.4	67.8	154	309.2	98.9	210	410.0	398.9	750	1382
37.2	99	210.2	68.3	155	311.0	99.4	211	411.8	404.4	760	1400
37.8	100	212.0	68.9	156	312.8	100.0	212	413.6	410.0	770	1418
38.3	101	213.8	69.4	157	314.6	104.4	220	428.0	415.6	780	1436
38.9	102	215.6	70.0	158	316.4	110.0	230	446.0	421.1	790	1454
39.4	103	217.4	70.6	159	318.2	115.6	240	464.0	426.7	800	1472
40.0	104	219.2	71.1	160	320.0	121.1	250	482.0	432.2	810	1490
40.6	105	221.0	71.7	161	321.8	126.7	260	500.0	437.8	820	1508
41.1	106	222.8	72.2	162	323.6	132.2	270	518.0	443.3	830	1526
41.7	107	224.6	72.8	163	325.4	137.8	280	536.0	448.9	840	1544
42.2	108	226.4	73.3	164	327.2	143.3	290	554.0	454.4	850	1562
42.8	109	228.2	73.9	165	329.0	148.9	300	572.0	460.0	860	1580
43.3	110	230.0	74.4	166	330.8	154.4	310	590.0	465.6	870	1598
43.9	111	231.8	75.0	167	332.6	160.0	320	608.0	471.1	880	1616
44.4	112	233.6	75.6	168	334.4	165.6	330	626.0	476.7	890	1634
45.0	113	235.4	76.1	169	336.2	171.1	340	644.0	482.2	900	1652
45.6	114	237.2	76.7	170	338.0	176.7	350	662.0	487.8	910	1670
46.1	115	239.0	77.2	171	339.8	182.2	360	680.0	493.3	920	1688
46.7	116	240.8	77.8	172	341.6	187.8	370	698.0	498.9	930	1706
47.2	117	242.6	78.3	173	343.4	193.3	380	716.0	504.4	940	1724
47.8	118	244.4	78.9	174	345.2	198.9	390	734.0	510.0	950	1742
48.3	119	246.2	79.4	175	347.0	204.4	400	752.0	515.6	960	1760
48.9	120	248.0	80.0	176	348.8	210	410	770.0	521.1	970	1778
49.4	121	249.8	80.6	177	350.6	215.6	420	788	526.7	980	1796
50.0	122	251.6	81.1	178	352.4	221.1	430	806	532.2	990	1814
50.6	123	253.4	81.7	179	354.2	226.7	440	824	537.8	1000	1832
51.1	124	255.2	82.2	180	356.0	232.2	450	842	565.6	1050	1922
51.7	125	257.0	82.8	181	357.8	237.8	460	860	593.3	1100	2012
52.2	126	258.8	83.3	182	359.6	243.3	470	878	621.1	1150	2102
52.8	127	260.6	83.9	183	361.4	248.9	480	896	648.9	1200	2192
53.3	128	262.4	84.4	184	363.2	254.4	490	914	676.7	1250	2282
53.9	129	264.2	85.0	185	365.0	260.0	500	932	704.4	1300	2372
54.4	130	266.0	85.6	186	366.8	265.6	510	950	732.2	1350	2462
55.0	131	267.8	86.1	187	368.6	271.1	520	968	760.0	1400	2552
55.6	132	269.6	86.7	188	370.4	276.7	530	986	787.8	1450	2642
56.1	133	271.4	87.2	189	372.2	282.2	540	1004	815.6	1500	2732
56.7	134	273.2	87.8	190	374.0	287.8	550	1022	1093.9	2000	3632
57.2	135	275.0	88.3	191	375.8	293.3	560	1040	1648.9	3000	5432
57.8	136	276.8	88.9	192	377.6	298.9	570	1058	2760.0	5000	9032

Above 1000 in the center column, the table increases in increments of 50. To convert 1462 degrees F. to Celsius, for instance, add to the Celsius equivalent of 1400 degrees F. ten times the interpolation factor for 6 and the interpolation factor for 2 or 760.0 + 33.3 + 1.11, which equals 794.4.

Conversion Tables of Angular Measure. — The accompanying tables of degrees, minutes, and seconds into radians; radians into degrees, minutes, and seconds; radians into degrees and decimals of a degree; and minutes and seconds into decimals of a degree and vice versa facilitate the quick conversion of angular measurements.

Example: The Degrees, Minutes, and Seconds into Radians Table is used to find the number of radians in 324 degrees, 25 minutes, 13 seconds as follows:

300 degrees	= 5.235988 radians
20 degrees	= 0.349066 radian
4 degrees	= 0.069813 radian
25 minutes	= 0.007272 radian
13 seconds	= 0.000063 radian
$324°25'13''$	= 5.662202 radians

Degrees, Minutes, and Seconds into Radians
(Based on 180 degrees = π radians)

Degrees into Radians											
Deg.	Rad.	Deg.	Rad.	Deg.	Rad.	Deg.	Rad.	Deg.	Rad.	Deg.	Rad.
1000	17.453293	100	1.745329	10	0.174533	1	0.017453	0.1	0.001745	0.01	0.000175
2000	34.906585	200	3.490659	20	0.349066	2	0.034907	0.2	0.003491	0.02	0.000349
3000	52.359878	300	5.235988	30	0.523599	3	0.052360	0.3	0.005236	0.03	0.000524
4000	69.813170	400	6.981317	40	0.698132	4	0.069813	0.4	0.006981	0.04	0.000698
5000	87.266463	500	8.726646	50	0.872665	5	0.087266	0.5	0.008727	0.05	0.000873
6000	104.719755	600	10.471976	60	1.047198	6	0.104720	0.6	0.010472	0.06	0.001047
7000	122.173048	700	12.217305	70	1.221730	7	0.122173	0.7	0.012217	0.07	0.001222
8000	139.626340	800	13.962634	80	1.396263	8	0.139626	0.8	0.013963	0.08	0.001396
9000	157.079633	900	15.707963	90	1.570796	9	0.157080	0.9	0.015708	0.09	0.001571
10000	174.532925	1000	17.453293	100	1.745329	10	0.174533	1.0	0.017453	0.10	0.001745

Minutes into Radians											
Min.	Rad.	Min.	Rad.	Min.	Rad.	Min.	Rad.	Min.	Rad.	Min.	Rad.
1	0.000291	11	0.003200	21	0.006109	31	0.009018	41	0.011926	51	0.014835
2	0.000582	12	0.003491	22	0.006400	32	0.009308	42	0.012217	52	0.015126
3	0.000873	13	0.003782	23	0.006690	33	0.009599	43	0.012508	53	0.015417
4	0.001164	14	0.004072	24	0.006981	34	0.009890	44	0.012799	54	0.015708
5	0.001454	15	0.004363	25	0.007272	35	0.010181	45	0.013090	55	0.015999
6	0.001745	16	0.004654	26	0.007563	36	0.010472	46	0.013381	56	0.016290
7	0.002036	17	0.004945	27	0.007854	37	0.010763	47	0.013672	57	0.016581
8	0.002327	18	0.005236	28	0.008145	38	0.011054	48	0.013963	58	0.016872
9	0.002618	19	0.005527	29	0.008436	39	0.011345	49	0.014254	59	0.017162
10	0.002909	20	0.005818	30	0.008727	40	0.011636	50	0.014544	60	0.017453

Seconds into Radians											
Sec.	Rad.	Sec.	Rad.	Sec.	Rad.	Sec.	Rad.	Sec.	Rad.	Sec.	Rad.
1	0.000005	11	0.000053	21	0.000102	31	0.000150	41	0.000199	51	0.000247
2	0.000010	12	0.000058	22	0.000107	32	0.000155	42	0.000204	52	0.000252
3	0.000015	13	0.000063	23	0.000112	33	0.000160	43	0.000208	53	0.000257
4	0.000019	14	0.000068	24	0.000116	34	0.000165	44	0.000213	54	0.000262
5	0.000024	15	0.000073	25	0.000121	35	0.000170	45	0.000218	55	0.000267
6	0.000029	16	0.000078	26	0.000126	36	0.000175	46	0.000223	56	0.000271
7	0.000034	17	0.000082	27	0.000131	37	0.000179	47	0.000228	57	0.000276
8	0.000039	18	0.000087	28	0.000136	38	0.000184	48	0.000233	58	0.000281
9	0.000044	19	0.000092	29	0.000141	39	0.000189	49	0.000238	59	0.000286
10	0.000048	20	0.000097	30	0.000145	40	0.000194	50	0.000242	60	0.000291

Radians into Degrees and Decimals of a Degree

(Based on π radians = 180 degrees)

Rad.	Deg.	Rad.	Deg.	Rad.	Deg.	Rad.	Deg.	Rad.	Deg.	Rad.	Deg.
10	572.9578	1	57.2958	0.1	5.7296	0.01	0.5730	0.001	0.0573	0.0001	0.0057
20	1145.9156	2	114.5916	0.2	11.4592	0.02	1.1459	0.002	0.1146	0.0002	0.0115
30	1718.8734	3	171.8873	0.3	17.1887	0.03	1.7189	0.003	0.1719	0.0003	0.0172
40	2291.8312	4	229.1831	0.4	22.9183	0.04	2.2918	0.004	0.2292	0.0004	0.0229
50	2864.7890	5	286.4789	0.5	28.6479	0.05	2.8648	0.005	0.2865	0.0005	0.0286
60	3437.7468	6	343.7747	0.6	34.3775	0.06	3.4377	0.006	0.3438	0.0006	0.0344
70	4010.7046	7	401.0705	0.7	40.1070	0.07	4.0107	0.007	0.4011	0.0007	0.0401
80	4583.6624	8	458.3662	0.8	45.8366	0.08	4.5837	0.008	0.4584	0.0008	0.0458
90	5156.6202	9	515.6620	0.9	51.5662	0.09	5.1566	0.009	0.5157	0.0009	0.0516
100	5729.5780	10	572.9578	1.0	57.2958	0.10	5.7296	0.010	0.5730	0.0010	0.0573

Radians into Degrees, Minutes and Seconds

(Based on π radians = 180 degrees)

Rad.	Angle	Rad.	Angle	Rad.	Angle	Rad.	Angle	Rad.	Angle	Rad.	Angle
10	572°57′28″	1	57°17′45″	0.1	5°43′46″	0.01	0°34′23″	0.001	0°3′26″	0.0001	0°0′21″
20	1145°54′56″	2	114°35′30″	0.2	11°27′33″	0.02	1°8′45″	0.002	0°6′53″	0.0002	0°0′41″
30	1718°52′24″	3	171°53′14″	0.3	17°11′19″	0.03	1°43′8″	0.003	0°10′19″	0.0003	0°1′2″
40	2291°49′52″	4	229°10′59″	0.4	22°55′6″	0.04	2°17′31″	0.004	0°13′45″	0.0004	0°1′23″
50	2864°47′20″	5	286°28′44″	0.5	28°38′52″	0.05	2°51′53″	0.005	0°17′11″	0.0005	0°1′43″
60	3437°44′48″	6	343°46′29″	0.6	34°22′39″	0.06	3°26′16″	0.006	0°20′38″	0.0006	0°2′4″
70	4010°42′16″	7	401°4′14″	0.7	40°6′25″	0.07	4°0′39″	0.007	0°24′4″	0.0007	0°2′24″
80	4583°39′44″	8	458°21′58″	0.8	45°50′12″	0.08	4°35′1″	0.008	0°27′30″	0.0008	0°2′45″
90	5156°37′13″	9	515°39′43″	0.9	51°33′58″	0.09	5°9′24″	0.009	0°30′56″	0.0009	0°3′6″
100	5729°34′41″	10	572°57′28″	1.0	57°17′45″	0.10	5°43′46″	0.010	0°34′23″	0.0010	0°3′26″

Minutes and Seconds into Decimals of a Degree and Vice Versa

(Based on 1 second = 0.00027778 degrees)

Minutes into Decimals of a Degree						Seconds into Decimals of a Degree					
Min.	Deg.	Min.	Deg.	Min.	Deg.	Sec.	Deg.	Sec.	Deg.	Sec.	Deg.
1	0.0167	21	0.3500	41	0.6833	1	0.0003	21	0.0058	41	0.0114
2	0.0333	22	0.3667	42	0.7000	2	0.0006	22	0.0061	42	0.0117
3	0.0500	23	0.3833	43	0.7167	3	0.0008	23	0.0064	43	0.0119
4	0.0667	24	0.4000	44	0.7333	4	0.0011	24	0.0067	44	0.0122
5	0.0833	25	0.4167	45	0.7500	5	0.0014	25	0.0069	45	0.0125
6	0.1000	26	0.4333	46	0.7667	6	0.0017	26	0.0072	46	0.0128
7	0.1167	27	0.4500	47	0.7833	7	0.0019	27	0.0075	47	0.0131
8	0.1333	28	0.4667	48	0.8000	8	0.0022	28	0.0078	48	0.0133
9	0.1500	29	0.4833	49	0.8167	9	0.0025	29	0.0081	49	0.0136
10	0.1667	30	0.5000	50	0.8333	10	0.0028	30	0.0083	50	0.0139
11	0.1833	31	0.5167	51	0.8500	11	0.0031	31	0.0086	51	0.0142
12	0.2000	32	0.5333	52	0.8667	12	0.0033	32	0.0089	52	0.0144
13	0.2167	33	0.5500	53	0.8833	13	0.0036	33	0.0092	53	0.0147
14	0.2333	34	0.5667	54	0.9000	14	0.0039	34	0.0094	54	0.0150
15	0.2500	35	0.5833	55	0.9167	15	0.0042	35	0.0097	55	0.0153
16	0.2667	36	0.6000	56	0.9333	16	0.0044	36	0.0100	56	0.0156
17	0.2833	37	0.6167	57	0.9500	17	0.0047	37	0.0103	57	0.0158
18	0.3000	38	0.6333	58	0.9667	18	0.0050	38	0.0106	58	0.0161
19	0.3167	39	0.6500	59	0.9833	19	0.0053	39	0.0108	59	0.0164
20	0.3333	40	0.6667	60	1	20	0.0056	40	0.0111	60	0.0167

Example 1: Convert 11′37″ to decimals of a degree. From the left table, 11′ = 0.1833 degree. From the right table, 37″ = 0.0103 degree. Adding, 11′37″ = 0.1833 + 0.0103 = 0.1936 degree.

Example 2: Convert 0.1234 degree to minutes and seconds. From the left table, 0.1167 degree = 7′. Subtracting 0.1167 from 0.1234 gives 0.0067. From the right table, 0.0067 = 24″ so that 0.1234 = 7′24″.

Revolutions per Minute for Various Cutting Speeds and Diameters

dia., mm	Cutting Speed, Metres per Minute									
	5	6	8	10	12	16	20	25	30	35
	Revolutions per Minute									
5	318	382	509	637	764	1019	1273	1592	1910	2228
6	265	318	424	530	637	849	1061	1326	1592	1857
8	199	239	318	398	477	637	796	995	1194	1393
10	159	191	255	318	382	509	637	796	955	1114
12	133	159	212	265	318	424	531	663	796	928
16	99.5	119	159	199	239	318	398	497	597	696
20	79.6	95.5	127	159	191	255	318	398	477	557
25	63.7	76.4	102	127	153	204	255	318	382	446
30	53.1	63.7	84.9	106	127	170	212	265	318	371
35	45.5	54.6	72.8	90.9	109	145	182	227	273	318
40	39.8	47.7	63.7	79.6	95.5	127	159	199	239	279
45	35.4	42.4	56.6	70.7	84.9	113	141	177	212	248
50	31.8	38.2	51	63.7	76.4	102	127	159	191	223
55	28.9	34.7	46.3	57.9	69.4	92.6	116	145	174	203
60	26.6	31.8	42.4	53.1	63.7	84.9	106	133	159	186
65	24.5	29.4	39.2	49	58.8	78.4	98	122	147	171
70	22.7	27.3	36.4	45.5	54.6	72.8	90.9	114	136	159
75	21.2	25.5	34	42.4	51	68	84.9	106	127	149
80	19.9	23.9	31.8	39.8	47.7	63.7	79.6	99.5	119	139
90	17.7	21.2	28.3	35.4	42.4	56.6	70.7	88.4	106	124
100	15.9	19.1	25.5	31.8	38.2	51	63.7	79.6	95.5	111
110	14.5	17.4	23.1	28.9	34.7	46.2	57.9	72.3	86.8	101
120	13.3	15.9	21.2	26.5	31.8	42.4	53.1	66.3	79.6	92.8
130	12.2	14.7	19.6	24.5	29.4	39.2	49	61.2	73.4	85.7
140	11.4	13.6	18.2	22.7	27.3	36.4	45.5	56.8	68.2	79.6
150	10.6	12.7	17	21.2	25.5	34	42.4	53.1	63.7	74.3
160	9.9	11.9	15.9	19.9	23.9	31.8	39.8	49.7	59.7	69.6
170	9.4	11.2	15	18.7	22.5	30	37.4	46.8	56.2	65.5
180	8.8	10.6	14.1	17.7	21.2	28.3	35.4	44.2	53.1	61.9
190	8.3	10	13.4	16.8	20.1	26.8	33.5	41.9	50.3	58.6
200	8	9.5	12.7	15.9	19.1	25.5	31.8	39.8	47.7	55.7
220	7.2	8.7	11.6	14.5	17.4	23.1	28.9	36.2	43.4	50.6
240	6.6	8	10.6	13.3	15.9	21.2	26.5	33.2	39.8	46.4
260	6.1	7.3	9.8	12.2	14.7	19.6	24.5	30.6	36.7	42.8
280	5.7	6.8	9.1	11.4	13.6	18.2	22.7	28.4	34.1	39.8
300	5.3	6.4	8.5	10.6	12.7	17	21.2	26.5	31.8	37.1
350	4.5	5.4	7.3	9.1	10.9	14.6	18.2	22.7	27.3	31.8
400	4	4.8	6.4	8	9.5	12.7	15.9	19.9	23.9	27.9
450	3.5	4.2	5.7	7.1	8.5	11.3	14.1	17.7	21.2	24.8
500	3.2	3.8	5.1	6.4	7.6	10.2	12.7	15.9	19.1	22.3

Revolutions per Minute for Various Cutting Speeds and Diameters (*Continued*)

dia., mm	Cutting Speed, Metres per Minute									
	40	45	50	55	60	65	70	75	80	85
	Revolutions per Minute									
5	2546	2865	3183	3501	3820	4138	4456	4775	5093	5411
6	2122	2387	2653	2918	3183	3448	3714	3979	4244	4509
8	1592	1790	1989	2188	2387	2586	2785	2984	3183	3382
10	1273	1432	1592	1751	1910	2069	2228	2387	2546	2706
12	1061	1194	1326	1459	1592	1724	1857	1989	2122	2255
16	796	895	995	1094	1194	1293	1393	1492	1591	1691
20	637	716	796	875	955	1034	1114	1194	1273	1353
25	509	573	637	700	764	828	891	955	1019	1082
30	424	477	530	584	637	690	743	796	849	902
35	364	409	455	500	546	591	637	682	728	773
40	318	358	398	438	477	517	557	597	637	676
45	283	318	354	389	424	460	495	531	566	601
50	255	286	318	350	382	414	446	477	509	541
55	231	260	289	318	347	376	405	434	463	492
60	212	239	265	292	318	345	371	398	424	451
65	196	220	245	269	294	318	343	367	392	416
70	182	205	227	250	273	296	318	341	364	387
75	170	191	212	233	255	276	297	318	340	361
80	159	179	199	219	239	259	279	298	318	338
90	141	159	177	195	212	230	248	265	283	301
100	127	143	159	175	191	207	223	239	255	271
110	116	130	145	159	174	188	203	217	231	246
120	106	119	133	146	159	172	186	199	212	225
130	97.9	110	122	135	147	159	171	184	196	208
140	90.9	102	114	125	136	148	159	171	182	193
150	84.9	95.5	106	117	127	138	149	159	170	180
160	79.6	89.5	99.5	109	119	129	139	149	159	169
170	74.9	84.2	93.6	103	112	122	131	140	150	159
180	70.7	79.6	88.4	97.3	106	115	124	133	141	150
190	67	75.4	83.8	92.1	101	109	117	126	134	142
200	63.7	71.6	79.6	87.5	95.5	103	111	119	127	135
220	57.9	65.1	72.3	79.6	86.8	94	101	109	116	123
240	53.1	59.7	66.3	72.9	79.6	86.2	92.8	99.5	106	113
260	49	55.1	61.2	67.3	73.4	79.6	85.7	91.8	97.9	104
280	45.5	51.1	56.8	62.5	68.2	73.9	79.6	85.3	90.9	96.6
300	42.4	47.7	53.1	58.3	63.7	69	74.3	79.6	84.9	90.2
350	36.4	40.9	45.5	50	54.6	59.1	63.7	68.2	72.8	77.3
400	31.8	35.8	39.8	43.8	47.7	51.7	55.7	59.7	63.7	67.6
450	28.3	31.8	35.4	38.9	42.4	46	49.5	53.1	56.6	60.1
500	25.5	28.6	31.8	35	38.2	41.4	44.6	47.7	50.9	54.1